T0229558

The Electronics Industry Research Series

- ■ The Taiwan Electronics Industry
- ■ The Singapore and Malaysia Electronics Industries
- ■ The Korean Electronics Industry

THE Korean

ELECTRONICS INDUSTRIES

Michael Pecht

J.B. Bernstein

D. Searls

M. Peckerar

Pramod C. Karulkar

CRC Press
Taylor & Francis Group
Boca Raton London New York

CRC Press is an imprint of the
Taylor & Francis Group, an **informa** business

CRC Press
Taylor & Francis Group
6000 Broken Sound Parkway NW, Suite 300
Boca Raton, FL 33487-2742

First issued in hardback 2018

© 1997 by Taylor & Francis Group, LLC
CRC Press is an imprint of Taylor & Francis Group, an Informa business

No claim to original U.S. Government works

ISBN 13: 978-1-138-43466-0 (hbk)
ISBN 13: 978-0-8493-3172-5 (pbk)

Visit the Taylor & Francis Web site at
http://www.taylorandfrancis.com

and the CRC Press Web site at
http://www.crcpress.com

CONTENTS

PREFACE

Since the Korean War, Korea has made substantial gains in building a modern economy and society. One area where the results of four decades of work is especially apparent is in the country's electronics industry. Today, Korean electronics companies own a significant share of the world electronics market. They conduct state-of-the-art research and development projects, establish foreign ventures, and support universities and research institutes devoted to producing the country's future engineers, scientists, and technical leaders.

Financial data and some general information on Korean electronics companies can be found in the company annual reports, on the Internet, and in Asian or Korean business publications. But a clear English-language analysis does not exist of how the Korean corporations operate, how employees view their own companies, and what makes them so financially successful. This book, based on visits to Korean electronics companies, conversations with a small number of company employees, and research of the available literature, attempts to answer some of these questions.

Under the direction of the U.S. World Technology Evaluation Center (WTEC), a panel of electronics experts with diverse backgrounds were selected to evaluate the multiple facets of electronics in Korea. As a result of WTEC organization, the panel visited various companies, institutions, and government agencies in Korea. This study aimed at publicizing only information that our hosts in Korea were willing to place in the public domain. In order to make perfectly certain that nothing confidential was published, each host was given the opportunity to review a draft of the report before it was published. This allowed them to delete sensitive material and/or correct errors. We did not look at North Korea, so any reference to Korea will mean exclusively The Republic of Korea (i.e., South Korea). Nevertheless, it is interesting to note that the whole time the panel was in South Korea, hardly a day went by when North Korea was not mentioned in the news, generally in terms of a strengthening military position along the border. The population lives with a constant threat of attack but appears not to take these threats seriously. In fact, those Koreans we met with commonly believed that they will be reunited with the North by the turn of the century. Nonetheless, the threat of aggression from the North looms as a constant reminder that this is not a nation at peace with all of its neighbors.

This book documents the technologies, manufacturing procedures, capabilities, and infrastructure that have made Korea so successful in the electronics industry. This knowledge, coupled with understanding of the future direction of the Korean electronics industry, is vital for U.S. competitiveness. Such information is needed to determine in which market sectors to compete and in which areas subcontracting, outsourcing, and partnership agreements would be beneficial.

The book consists of seven chapters. Chapter 1 presents an overview of the industrial setting in which Korea's electronics industry operates, including discussion of the roles of the government and universities in establishing an infrastructure in which the Korean electronics industry can flourish. Chapters 2, 3, and 4 present specific information about the major segments of the Korean electronic industry: semiconductors; displays; and packaging, printed circuit boards, and systems. Chapters 5 and 6 contain descriptions of key electronics companies and institutes and universities supporting electronics in Korea, many of which were visited by the study team. Chapter 7 gives an integrated view of the panel's findings and impressions

A WTEC-sponsored workshop was held on July 18, 1996, in Washington DC, at which the panel presented preliminary findings. All Korean hosts and selected Korean dignitaries were invited to participate. In addition, representatives from U.S. government agencies, electronics companies, universities, and other interested parties attended. A brief biography of each of the panel members is given below.

Michael G. Pecht (*Panel Chair*) is a Professor and Director of the CALCE Electronic Packaging Research Center at the University of Maryland. He is an IEEE Fellow, an ASME Fellow, and a Westinghouse Fellow. He is chief editor of the IEEE *Transactions on Reliability* and the author of eleven books on electronic packaging.

Barbara J. Shula is Director of Quality and Reliability Engineering at Advanced Micro Devices (AMD) and is responsible for their worldwide manufacturing operations. She has worked at Hewlett Packard, Lanax, and IBM.

John Budnaitis is a principal engineer at W. L. Gore & Associates specializing in wafer-scale testing and burn-in testing. Prior to his present position he worked for Supercomputer Systems, Micron, and Intel.

Joseph B. Bernstein is a faculty member in the Reliability Engineering Program at the University of Maryland. Prior to his present position he worked for MIT Lincoln Laboratories specializing in electronic manufacturing processes and defect avoidance.

Carl A. Rust was responsible for technology transfer, industry relations, research strategy, operations, and new business development at the CALCE Electronic Packaging Research Center. He spent seven years working in Quality and Reliability Engineering at Texas Instruments and is now at Georgia Tech.

Martin Peckerar is head of the Thin Film and Interface Sciences Branch of the Naval Research Laboratory (NRL). He has worked for Westinghouse as a group leader of advanced MOS process development and as a professor of electrical engineering. He is a Fellow of the IEEE and coauthor of the textbook *Electronic Materials: Science and Technology.*

Pramode C. Karulkar is Manager of the Microelectronics Research Laboratory for the University Research Foundation (URF). He has worked at MIT Lincoln Laboratory, the Microelectronics Manufacturing Division of Hughes, and Custom Device Manufacturing Group of Rockwell.

Mary J. Li is a research scientist at the CALCE Electronic Packaging Research Center. Her expertise is in the areas of materials characterization and failure analysis of electronics.

The authors of this book are deeply indebted to the large number of people who contributed to its preparation. We extend our sincere appreciation to the panel and our Korean hosts who so graciously gave of their time in sharing information about their respective organizations. We are particularly grateful to the Korean Ministry of Science and Technology for helping coordinate our visit. We also thank Ken Flamm of the Brookings Institution (Foreign Policy Studies Program); Dr. Betty Prince of Memory Strategies International; Matt Doty of Amkor Electronics Inc.; O.J. Kwon of Kyungpook National University; Jong Kim of CALCE EPRC; Daniel Lee, Manager of Corporate Communications, LG Electronics, USA; and Sharon Yun of the Office of Technology Policy, Technology Administration, U.S. Department of Commerce. Finally, thanks are due to the WTEC staff at Loyola College (Geoff Holdridge and Duane Shelton) and to the National Science Foundation (Paul Herer), which partially funded this study under its Cooperative Agreement with WTEC at Loyola College (ENG-9416970). The Government has certain rights in this material. Any opinions, findings, and conclusions or recommendations expressed in this material are those of the authors and do not necessarily reflect the views of the U.S. Government, Loyola College, or the authors' parent institutions.

Chapter 1

THE KOREAN ELECTRONICS INDUSTRY

The Republic of Korea's industrialization of its economy has been among the most rapid in the world, and the country has sustained strong economic growth for much of the last two decades. With a population of 45 million, Korea's 1995 Gross Domestic Product topped $447.6 billion, with a per capita GDP of $9,990 (Table 1.1). This per capita GDP is expected to top $10,000 given a continued growth rate over the next three years of over 6%.[DOC 1996] As another indicator of Korea's strong economic growth, Korea's manufacturing output increased by an annual average of 9.9% from 1979 to 1995. This striking growth has propelled the country to eleventh place among the world's top economic powers.

The rapid economic growth of the 1970s and early 1980s was tempered by a slowdown in the late 1980s that mirrored a variety of socioeconomic changes taking place in Korea. The most serious economic problems — a deficit in the current account balance, 10% inflation, and a dropping GDP growth rate — improved during the first half of the 1990s. While the GDP and wage rates continued to rise (wages rose 14.5% from 1994 to 1996), inflation held steady at 6%, and Korea's foreign debt remained low (at $60.5 million in 1995). These positive economic indicators reflect both Korea's continued economic stability and its increasing standard of living, now equivalent to that of Hong Kong, Singapore, and Taiwan. However, while many economic indicators are very positive, some underlying factors continue to put pressure on the Korean economy, including rising wage rates, which contribute to Korea's growing trade deficit and declining share of the light industry manufacturing market.

Table 1.1
Korea's Economic Growth, 1992-1996 (in $ billions)

	1992	1993	1994	1995	1996 (projected)
GDP	307.9	332.8	379.5	447.6	480
Per Capita	7,007	7,513	8,483	9,990	10,000+

Source: DOC 1996

1.1 KOREAN INDUSTRY

1.1.1 Korea's Industrial Development Stages

Korea's industrial and competitive development has advanced through well-defined stages to reach its present position (Fig. 1.1). This progression reflects the development of Korea's industrial economy in general and the development of its competitive advantages in particular. It also indicates the government's role in the development of a science and technology (S&T) infrastructure in Korea.

During Korea's "factor-driven" development stage (1960s and 1970s), the country's key competitive advantages were its abundant and inexpensive labor and its ability to acquire technology through the purchase and imitation of foreign capital goods. Korea capitalized on these competitive advantages by expanding its export-oriented light industry, investing in large-scale production facilities, and improving its process technology. At this time, the government began taking a leadership role in developing Korea's technological competitiveness by establishing scientific institutions and foundations for its scientific infrastructure. It also launched national R&D programs to promote assimilation of imported foreign technology.

Korea is now in an "investment-driven" development stage where it has concentrated for the last two decades on acquiring and assimilating the most advanced technology available in the global marketplace. As its manufacturing capability improved, this became Korea's competitive advantage, enabling it to expand its technology-intensive industries and increase R&D capacity. The investment stage has been characterized by a pragmatic, integrated strategy that harnesses the strengths of the private sector, academia, and government. Together, the public and private sectors of Korea have strategically focused on maintaining and increasing competitiveness by increasing technological self-sufficiency, identifying new sources of technology in such places as Russia and China, developing new products from technological sources other than traditional suppliers, and increasing foreign investment by Korean firms, especially in Southeast Asia, the United States, and Europe. Technology transfer and assimilation have formed the basis of the Korean competitive infrastructure.

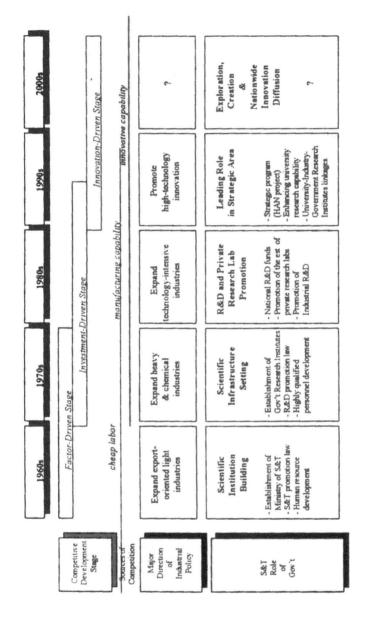

Figure 1.1. Evolution of competitive development stages and government role in moving Korea's economy to innovation stage. [DOC 1996]

The Korean government has retained its leadership role, channeling limited capital into specific industries, promoting risk-taking, providing protection of its domestic markets to encourage the entry of national firms and the construction of efficient facilities, stimulating and influencing the acquisition of foreign technology, and encouraging exports. The government continues to offer a number of incentives to local industries to expand their investments in R&D, including accelerated depreciation allowances, investment tax credits, deferral of income tax payments, and duty-free import of selected capital goods.

The government is also involved with business development, supporting strategic industries, and it has significantly boosted R&D spending to develop key technologies aimed at increasing competitiveness in global markets. A notable example is the planned investment of some $27 million for the research and development of High Definition Television (HDTV) technology. The government is also active in creating and promoting a nationwide science culture within its society, with the cooperation of the academic, industrial, and media communities. This concerted public support for science and education has provided a strong foundation for individual and community support for technological infrastructure development.

The Korean government has placed a high priority on developing Korea's own indigenous technological capabilities and creating a world-class industrial infrastructure by the year 2001. It has recognized that economies with a strong domestic technological infrastructure are better equipped to play the game of global competition. Korea continues to parlay its dominance as a processor rather than an innovator and is still gathering and assimilating knowledge from outside sources to build technological strength. The government and "chaebols" (industrial conglomerates) have worked both independently and together to make up for the shortcomings of their technological infrastructure and are focusing on obtaining foreign technology from industrialized nations through direct and indirect methods.

Korea is slowly making the transition from the "investment-driven" development stage to the "innovation-driven" stage in which its major source of competitive advantage will be innovation capability. This transition is fraught with challenges, the most difficult being the generation of high rates of technological innovation. Many of Korea's chaebols are already competing in the sophisticated global market, and consequently, Korea is undergoing structural changes that favor high-value-added industries. Leading chaebols such as Samsung, LG, Hyundai, and Daewoo are positioned at the forefront of their industries and are in the process of globalizing their R&D centers to develop state-of-the-art technologies. However, while these chaebols have begun to support technological innovation, Korea is still weak in terms of creating innovative scientific knowledge and technology.

1.1.2 Chaebols

Korea's economy is dominated by a relatively small number of "chaebols," which are large industrial companies, usually owned and controlled by a family, that often enjoy vertical monopolies (i.e., the company and its subsidiaries control most of the steps in production, from the acquisition of the raw materials to fabrication), and that extend horizontally across diverse industries. In the immediate post-Korean War years, the chaebols took advantage of subsidized loans and tax breaks provided by the Korean government that allowed them to grow very quickly and swallow up smaller businesses. By the early 1980s, the four most prominent chaebols, Samsung, LG, Hyundai, and Daewoo, were immense.

In recent years, the government has taken steps to restrict the chaebols' loan access, thereby encouraging the establishment and growth of small- to medium-sized companies. In addition, family members who dominated the top management of these conglomerates have largely been replaced by executives with international experience.

One strategy of the major Korean chaebols is their diversification or horizontal integration strategy, similar to that of the Japanese *keiretsu*. As they dominate in a single market, they develop sectors in many unrelated markets as well. The major Korean chaebols produce cars, appliances, and semiconductor components, among many other things. This strategy protects companies as well as employees from seasonal downturns in one industry.

1.1.3 Corporate Culture

Korea's corporate culture can be difficult for U.S. business people to understand. Nationalism plays a key role: among professionals and workers alike, there is a common resolve for Korea to become an economic giant in the world market. Also, in the observation of the WTEC panelists, business groups appear to be like enormous families, where the growth of the whole group is targeted but is accomplished through interdependent growth of smaller units at successively lower levels down to the individual employee. Companies enjoy loyalty from the employees, and companies return that loyalty. A man's decision to join a company — if able to pass the company exam — is influenced by mentors such as university professors who have contacts with the company or family members who work for the company. Once on board, employees, especially male employees, tend to stay with the company lifelong, although this is starting to change in the larger chaebols.

Korean companies often provide for all basic needs of blue-collar employees, providing housing, medical care, and recreation, as well as jobs, taking quite a "holistic" approach to worker compensation. Employees, in return, work with dedication and perseverance. Workers are unionized across the industry, so wages are basically fixed for a certain level of education and experience. Even when a company is enormously successful, workers tend not to argue for higher wages, which could hurt their companies' — and their country's — profitability and future prosperity.

In return for employee loyalty to their companies, companies appear to show great loyalty to employees. Upper management tends to heavily weigh effects on employee morale in their corporate decision-making processes. Workers' feelings and the sense of community are of utmost importance to everyone, as is the welfare of the nation as a whole. The personal respect that companies uniformly show to their employees makes everyone feel individually responsible for the success of both country and company.

Panel authors were unable to determine the role played by women in Korea's electronics industry. A large number of operators on the production lines are women, but the panel saw and met no women engineers during its visits to Korean companies.

Besides patriotism and mutual management-employee loyalty and teamwork, other standout characteristics of Korean corporate culture include highly focused and extremely hard work, manufacturing excellence, investment of a high percentage (10% or more) of revenues into R&D, and aggressive globalization.

1.1.4 Physical Infrastructure

One factor that may hamper further development of Korea's electronics industry is the country's physical infrastructure. Companies are growing at a pace that cannot be matched by the government's ability to build and improve roadways. Seoul's streets are so congested that a mid-day two-mile taxi ride can take over half an hour. While Korea's transportation system is very good for a developing nation, and Seoul has an extremely advanced subway system, the roadways must be improved as the major manufacturers expand. One way Korea's expanding companies have compensated for their country's lagging infrastructure is to build manufacturing plants in the countries in which they do business. For Korea, this strategy engenders good will with its trading partners by employing a domestic labor force, and it avoids some trade impediments and increases access to new technologies.

1.1.5 Foreign Trade

Korea's manufacturing industries have based much of their success on foreign trade. In 1995, Korea was ranked the world's 13th largest trading nation, with $96 billion in exports and $102 billion in imports. The United States has been Korea's most significant trading partner, although an annual trade surplus with the United States that reached $42 million at its zenith set off trade disputes in the late 1980s. Terms of settlement of those disputes contributed to the beginning of Korea's overall trade deficit, which totaled $1.2 billion in 1995.[DOC 1996] Exports continue to be essential to the stability and growth of the Korean economy. Among Korea's most profitable exports have been electronics goods. Korea expects to have a steady export growth of 12.3% in industrial electronics, 11.7% in electronic parts, and 4.7% in consumer electronics. [Kim 1994a] Table 1.2 shows the rising level of total electronic exports of Korea's leading firms.

Table 1.2
Value of Exports by Major Electronic Companies, 1987-1995

Companies	1987	1991	1993	1995
Samsung		$5 billion	$7 billion	$10 billion
LG	$1.8 billion	$2.1 billion	$2.9 billion	$5.0 billion
Hyundai	$390 million	$730 million	$1.1 billion	$4.1 billion
Daewoo	$800 million	$1.2 billion	$1.5 billion	$1.9 billion

Korean exports of semiconductors have been particularly profitable. Semiconductor exports totaled $10 billion in 1994, of which $4 billion was earned by Samsung Electronics Corporation. In roughly a decade, the Korean semiconductor companies have made a phenomenal impact on the world markets. Korean players in the semiconductor market are regarded by U.S. counterparts as serious competitors. They are technology implementors (e.g., through DRAM manufacturing); technology buyers (e.g., through acquisition of Japanese technology and of U.S. companies); and technology leaders (e.g., through development of a 1 Gbit DRAM).

1.2 THE ELECTRONICS INDUSTRY

In the late 1960s and early 1970s, Korea's electronics industry consisted mainly of assembly plants that manufactured common consumer electronics. Fundamental factors in the phenomenal success that the chaebols have built on this base include their heavy investment in improved manufacturing technology, thus allowing them to market competitively priced, high-quality products in large volume. In addition, Korean firms have become intensely involved in R&D activities, especially by means of joint ventures with foreign companies.

As an example of the evolution of Korean electronics companies, the LG Group started by assembling vacuum tube radios, transistor radios, tape recorders, and black-and-white televisions from imported components. It now has sales figures in the billions and production facilities in Asia, Europe, and North America, and it participates in a number of joint R&D projects with companies like AT&T, Sony, and IBM.

The major companies Samsung, LG, Hyundai, and Daewoo have committed to manufacturing consumer as well as industrial electronics goods. These companies export billions of dollars worth of electronic products and also devote millions of dollars to R&D activities. The payoff of these investments clearly can be seen in the corresponding growth of the country's economy and in the success stories of companies like Samsung, which only broke into the electronics market in 1982 but is now the world's largest DRAM manufacturer.

1.2.1 Korea's Electronics Products

The following lists the principal electronics products of Korea's largest four chaebols:

1.2.1.1 *Samsung*

- memory devices (DRAM, SRAM, video RAM, ROM, CARD, DRAM modules)
- audio & video (TVs, LCD projectors, components, CDP, LDP, CD-ROM)
- computer systems (mini computers, microcomputers, workstations, desktop/laptop PCs, network systems)
- optical filing systems
- micro devices:
    - discrete: MOS, linear, and logic IC; ASIC; DSP; microcomponents
    - telecommunication: TDX, MODEM, MUX, facsimile, copiers, typewriters, optical fibers, car phones, hand phones

1.2.1.2 *LG*

- audio & video equipment (TVs, VCRs, audio components)
- computer & office automation (desktop PCs, notebook PCs, color TFT-LCD displays, multisync monitors, CD-ROM drives, facsimiles, laser printers)
- multimedia (handheld PC, PDA, Cd-i)
- semiconductors (64 Mbit DRAM, MPACT media processor, RAMBUS DRAM, JAVA processor, digital signal processor, Gigabit DRAM, microcontrollers, application-specific IC products)

1.2.1.3 *Hyundai*

- semiconductor sector (memory, system IC, LCD)
- information sector (PCs, monitors, data communications, computer networks)
- industrial electronics sector
- telecommunications, new media

1.2.1.4 *Daewoo*

- CE capacitors, tantalum capacitors, film capacitors, DY, flyback transformers, E-tuners, hybrid ICs, saw filters, thermistors, keyboards, and printers

Table 1.3 gives the total sales of the top 4 companies between the years 1987 and 1995.

Table 1.3
Total Sales of Major Electronic Companies, 1987-1995

Companies	1987	1991	1993	1995
Samsung	$4 billion	$8 billion	$10 billion	$12 billion
LG	$3.5 billion	$3.8 billion	$5.5 billion	$7 billion
Hyundai	$0.6 billion	$1.1 billion	$1.6 billion	$5.1 billion
Daewoo	$1.1 billion	$2.0 billion	$2.5 billion	$3 billion

1.2.2 Imported vs. Domestic Technology Capabilities

A good deal of the technology crucial to the success of the Korean electronics firms historically has been transferred from other countries, primarily the United States and Japan. In general, Korea still favors these two countries as technology partners, but Korean companies are also pursuing technology in less obvious places, specifically Russia, the other former Soviet states, and China. These nations offer a fairly high level of education, a strong background in basic R&D, low prices for patented technology, and low labor rates. China also offers an advantageous geographic location that facilitates the flow of products and people, and cultural similarities that allow for ease of business communications. Korean companies look to these various countries for technology to directly enhance their economic growth and to indirectly help develop indigenous technologies to strengthen their global competitiveness into the 21st century.

Korea's electronics industry has done much since the 1970s to move away from its early, more passive approach to growth. Korean electronics capabilities have developed rapidly — the country now shares with Taiwan and Singapore almost 50% of the world market for manufacturing and assembly of electronic components. Although there is still heavy dependence in the larger companies on importing electronic parts and components, foreign-controlled production in Korea has been declining since the 1980s. Government institutes, universities, and industry are vigorously pursuing R&D, both independently and in joint ventures with foreign institutions, and they have a number of important innovations to their credit.

Largely as a means to be more self-reliant in technology innovation, joint venture companies have been on the rise in Korea, and this is a key Korean business strategy for remaining competitive in the global electronics market. Every major Korean company is involved in joint ventures, both to maintain the flow of technology with leading foreign companies and to distribute R&D costs. For example, LG Electronics has a joint venture company with Alps Electronics of Japan and has joint projects with Motorola to develop ASIC technology, with Xerox for Power PC research, with Philips for video CDs, and with Zenith for cable modem design and next-generation multimedia set-top box product concepts.

Cooperative agreements between Korean and foreign firms have long been a source of infrastructure and commercial technology for Korea. The chaebols generally bring cash, production process experience, and access to new markets with them to the negotiating table; foreign firms generally bring the necessary technology. Strategic technical alliances with other world market leaders allow the chaebols to penetrate new markets faster and give them access to a broader range of cutting-edge technology.

Korean companies continue to pay large royalties for foreign technology. In fact, royalty payments from January 1995 through February 1996 reached almost $1.9 billion — more than two-thirds of all of Korea's royalty payments, which totaled $2.5 billion to 1996. Electric and electronics industries made the largest share of the payments, and of these, Samsung's payments far exceeded the others.

Another important aspect of Korea's electronics industry is its dependence on original equipment manufacture (OEM) agreements, which allow Korean companies to sell their products under other companies' names. Such arrangements have been vital to international sales of Korean goods, since in the world electronics marketplace, products with recognized brand names tend to sell better than unknown brands, and consumer recognition of Korean firms' names is poor. Samsung noticed an appreciable increase in sales of its microwave ovens when it began selling them under the General Electric name. Taking advantage of the name-recognition factor, LG Electronics relies on NEC and Matsushita to sell its facsimile machines. However, Korean companies now observe that in order to lead the global electronics market with products bearing their own names, they must reduce exporting through OEM agreements, and most of them are striving to become independent of OEM agreements by increasing their prestige through innovations in design and technology.

1.3 THE SEMICONDUCTOR INDUSTRY

In the Korean semiconductor industry, the phenomenon is apparent — similar to that of the Japanese industry — of close government and industry cooperation in order to meet national goals. When the Korean government announced in the early 1980s that development of a semiconductor memory industry was a national priority, all major domestic companies dutifully expanded into the area. Samsung has had the most success, largely because the company already had experience with electronic components such as watch chips and medium-scale integrated circuits (ICs). Hyundai had a rockier start partly because of its lack of experience with chip manufacturing. It shut down its U.S. semiconductor operation in 1985; however, Hyundai's semiconductor business has grown steadily, and Hyundai Electronics America has invested $50 million in U.S. telecommunications, primarily through selling personal communications services (PCS) networks.

Japan and the United States dominated high-technology industries until the 1980s, when Korea started to compete in the DRAM market. In 1994 Samsung, LG, and Hyundai were in the global top-10 list of memory suppliers (Fig. 1.2). A major factor in this considerable success is that Korean manufacturers have successfully implemented more efficient mass production techniques, thus allowing for more competitive unit pricing. According to company statistics, total sales of the major four electronics companies grew more than 250% from 1993-1995 (Table 1.3). The Korean semiconductor industry captured nearly one-quarter of the world market in 1994, and specifically, Korea's share of the DRAM market increased from 10% in 1993 to a 1994 value of 29%.[Kim 1995]

In 1996, forecasts for Korean semiconductor exports have been revised downward several times due to a global market glut and plummeting prices for 16 Mbit DRAMs ($30 to $11). In response, the industry has pushed ahead with early transition to 64 Mbit DRAMs and higher-value-added memory chips.

As pressure has increased for manufacturers to develop more heavily integrated circuits in the sub-0.5 μm range, Korea's industries have faced a lack of adequate indigenous manufacturing technology. Much of the equipment relevant to microelectronic manufacturing, such as sputterers, ion implanters, and diffusion furnaces, are assembled by foreign equipment manufacturers. According to KSIA, domestic chip makers in 1994 relied on foreign manufacturers for 84% of equipment, 71% of devices, and 52% of materials. To counteract this dependence on foreign equipment manufacturers, the government began encouraging Korean companies to develop indigenous semiconductor manufacturing equipment through tax incentives and low-interest loans for R&D.

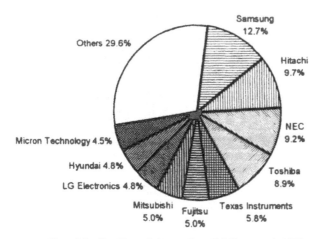

Figure 1.2.   Top 10 worldwide suppliers of MOS memory in 1994.

The dominance of Korean manufacturers in DRAM and other such commodity products is unquestionable; however, Korean industrialists recognize that the most sophisticated, highest-value-added semiconductor technologies are still dominated by the United States. They strive to emulate the largest U.S. companies, and they are improving their educational infrastructure, recruiting expertise, and domestic R&D facilities in the process.

1.3.1 Investments Overseas

Korean electronics chaebols are investing heavily in foreign ventures. LG Electronics purchased $350 million of stock in Zenith, the last TV manufacturing company in the United States, and LG Semiconductor signed a deal with Compass Design Automation to develop 0.35 μm tools. LG invested heavily in AT&T semiconductor activities starting in the 1980s. Samsung purchased 40% of the stock of AST Research for $378 million in February 1995 and agreed to invest another $138 million with Texas Instruments in a joint venture chip assembly plant in Portugal. Hyundai invested more than $1.3 billion to build its largest DRAM fabrication facility in Oregon for manufacture of 16 Mbit and 64 Mbit DRAMs. Expectations at the time of this report were that the plant would process more than 30,000 8-inch wafers per month and employ 1,000 workers. This investment was expected to bring Hyundai $350 million in 1996, $347 million in 1997, and $550 million in 1998. Even though these investments are small compared to Japanese investments in the United States, it is important to note that the rate of Korean overseas investing has been gradually increasing.

Another factor in overseas investments is rising domestic labor costs, which have forced some Korean companies to move manufacturing industries offshore to lower-wage countries like China and Indonesia.

1.3.2 Product Design for Foreign Markets

Korean companies are seeking to develop products specifically for the European market. A problem they face in this endeavor is to meet each country's requirements and specifications, which are often different from each other and from those of the United States. Samsung Semiconductor is one Korean company that is apparently adjusting well to this problem. It is enjoying exceptionally high growth in the European market (Table 1.4) compared to European, U.S., and Japanese firms.

Table 1.4
European Semiconductor Market Rankings

Company	Sales ($ million)		Growth
	1993	1994	1993/94
Intel	2056	2605	27%
Motorola	1260	1539	22%
Siemens	1032	1380	34%
Philips	1104	1269	15%
SGS-Thomson	977	1236	27%
Texas Instruments	871	1121	29%
Samsung	510	1007	98%
NEC	605	969	60%
Toshiba	577	772	34%
IBM Microelectronics	409	705	72%

Source: KETI 1994

1.4 THE ROLE OF GOVERNMENT IN S&T

Various Asian governments have strongly emphasized technology and innovation as a means of supporting competitiveness and stimulating their economies. The Korean government is one of these. It maintains an emphatically proactive position with respect to internal technology development that has been especially beneficial to the growth of its national semiconductor industry.

Government nurtures the electronics industry in Korea in two ways. First, it works to create a legislative basis for the growth of high-technology industry through tax incentives, banking laws, and laws of incorporation. Second, it charges a number of ministries with providing financial support to public and nonprofit institutes, universities, and other educational institutions. While individual programs and projects identified and sponsored by the legislature and the ministries may be criticized, their ultimate value is undeniable. The government programs lower the cost of basic research and plant modernization, and government funding of university and other educational programs improves the availability of skilled technologists and serves to maintain a pool of basic researchers and educators that serve as consultants to industry at large. Contrary to its planning of many other aspects of society, the Korean government is largely unintrusive into the workings of the major electronics corporations.

1.4.1 Outline for Technological Development

All pertinent activities of the Korean government are organized around a general outline for technological development, summarized in Table 1.5. This outline is broad in scope, but recognizes major domestic as well as global market trends. More specific definition of development is given by discipline. This output is the result of the joint government and industry study, known as the G7 study, which combines internal discussions of Korean industry trends with the roadmaps of extra-national agencies such as the U.S. SIA semiconductor development roadmap. Figure 1.1, shown earlier, indicates a similar assessment of the Korean government's long-term commitment to S&T development.[DOC 1996]

1.4.2 Legislative Interaction

As is true worldwide, tax incentives in Korea are a powerful tool for stimulating and maintaining high-profit industries. Korean companies may maintain a "reserve fund" for investment in technology development, including R&D, employee development, and plant improvement. Up to 4% of total sales may be applied to this fund and sheltered from taxes. In addition, companies may deduct 15% of total expenditures on training and manpower development. Ten percent of R&D facility construction costs are deductible, and R&D and test facilities may be depreciated at the rate of 90% per year. The Collaborative R&D Promotion Law of 1993 is aimed at nurturing interfirm collaboration through funding and tax advantages.

Table 1.5
General Outline for Korean Development

Period	Industrialization	S & T Development
1960s	Develop import substitution. Expand export-oriented light industries. Support producer goods industries.	Strengthen S&T education. Deepen S&T infrastructure. Promote foreign technology imports.
1970s	Expand heavy & chemical industries. Shift emphasis from capital import to technology import. Strengthen export-oriented industrial competitiveness.	Expand technical training. Improve institutional mechanism for adapting imported technology. Promote research applicable to industrial needs.
1980s	Transform industrial structure to one of comparative advantage. Expand technology-intensive industry. Encourage manpower development and improve industry productivity.	Develop and acquire top-level scientists and engineers. Launch the national R&D projects. Promote industrial technology development and industrial labs.
1990s	Promote industrial structure adjustment, technical innovation. Promote efficient use of human and other resources. Improve information networks.	Reinforce national R&D projects. Strengthen demand-oriented technology development system. Globalize R&D systems and information networks.

Government also "partners" with both established and emerging firms in new product development. Industries may engage in 50% cost-sharing of new technology development if their research overlaps goals of established national programs like the DRAM project, aimed at accelerating development of computer memory chip technology. The government will fund as much as 90% of new technology development by individuals or small firms.

Funds may be disbursed through relevant ministries or through government-owned corporations such as the Korean Electric Power Corporation (KEPCO). Banks such as Korea Development Bank, Citizens National Bank, Korea Technology Development Bank, and Industrial Bank of Korea provide low-interest loans for new product development and for commercialization of newly developed products. The Korea Technology Banking Corporation (KTB) supplies "venture capital" support for development activities.

The Korea Research and Development Information Center is setting up an "information superhighway" between it and ten regional development centers, allowing for rapid dissemination of new production methods and approaches to the country at large via Internet-type services. An effective industry standards activity is maintained at the Korea Research Institute of Standards and Science through the Industrial Advancement Administration, an affiliate of the Ministry of Trade, Industry, and Energy (MOTIE). Also, Korea is toughening its stand on intellectual property rights protection and is streamlining its government procurement system to aid domestic assimilation of new technology.

1.4.3 Government Sponsorship of R&D

The Korean government sponsors research in support of electronics technology primarily through (1) the Ministry of Science and Technology (MOST); (2) MOTIE; and (3) the Ministry of Information and Communication (MOIC). MOST supplies roughly $10 million of support annually to a variety of national research institutes. Table 1.6 lists some of those that have electronics programs. (For a more detailed description of KIST, see Chapter 6.) Some 10% of Korea's total science research budget goes to electronics; 75% of that goes to DRAM development.

1.4.4 Specific Government-Sponsored Electronics R&D Programs

1.4.4.1 DRAM Project

The DRAM project (now part of the Highly Advanced National, or HAN, project) is an example of direct government support of industry. For such projects as this, 50% cost-sharing is expected. Also, money accrued from licensing the technology derived from this research are expected to go back to the ministries for reinvestment. Table 1.7 documents expenditures in the DRAM project.

Table 1.6
MOST-Supported Institutes with Electronics Programs

Organization	Field of Research
Korea Institute of Science and Technology (KIST): Seoul	Core technology development
System Engineering Research Institute (SERI): Taejon	Software training and education on computer program
Science and Technology Policy Institute (STEPI): Seoul	Policy studies and evaluation of national R&D projects
Korea Research and Development Information Center (KORDIC): Taejon	R&D information activities
Korea Institute of Energy Research (KIER): Taejon	Development and utilization of energy technology
Korea Aerospace Research Institute (KARI): Taejon	Development of aerospace-related technology
Korean Research Institute of Standards & Science (KRISS): Taejon	National standards
Korea Electric Technology Research Institute (KETRI): Chaugwon	Development of technology pertaining to electric power

Table 1.7
Korean Research in ULSI Technology (4-256 Mbit DRAM)

	4 Mbit	16 Mbit	64 Mbit	256 Mbit
Research Period	'86.10 - '89.3	'89.4 - '91.3	'89.4 - '93.3	'93.11 - '97.11
R&D Budget (in $US millions)	$112: MOST $13 MOIC $25 Industry $74	$244: MOST $32, MOIC $32, MOTIE $13, Industry $167		
Patent Applications	154	343	308	100

1.4.4.2 HAN Projects

As part of it's globalization strategy, the Korean Government developed the Highly Advanced National projects to selectively develop key industrial technologies requiring nationwide R&D investment.. A HAN project is a large-scale R&D project carried out through joint investment by the government and the chaebols under a long-term project management system, which is supported by interministerial cooperation and coordination. Various R&D actors such as universities, industries, and government-supported research institutes participate in each project. For the areas where domestic R&D capacity is lacking, international cooperation and technology transfer are actively pursued.

Some of the technologies targeted by the HAN Projects are aerospace, automobiles, bioengineering, computers, communications, electronics, environment, machinery and metals, medical equipment, nuclear power, and semiconductors. The HAN program coordinates ULSI-related projects and assists in the development of broadband integrated services digital networks (B-ISDN), high-definition television (HDTV), and advanced manufacturing systems. Starting in 1995, the program began branching into such areas as advanced materials, micro-electromechanical systems, flat-panel displays, next-generation vehicle technology, environmental technology, and new energy technologies.

The HAN Project is a comprehensive program linking many disciplines and technologies together for synergistic advancement. The overall development plan was a product of the G-7 survey. Over its ten-year life (1992 through 2001), this program will spend $4.7 billion on the broad-ranging R&D programs shown in Table 1.8.

1.4.5 Government Support of Equipment Development

Industry welcomes government support in semiconductor equipment development, since Korea has no indigenous equipment supplier base. Lithography tools, vacuum deposition equipment, and CVD reactors must all be brought in from abroad. This creates at least a perception of vulnerability. It also means that Korean corporations have a minimal role in defining design-rule evolution.

To date, Korean chip manufacturers have expressed greater interest in working through international semiconductor associations (such as SEMI) or through direct contact with national associations (such as the U.S. SIA) than in working through the government to influence equipment development. But most companies interviewed indicate an interest in seeing some Korean initiatives in this area, with government support provided, especially through tax incentives. There is also some discussion of the government providing land in an industrial park environment for equipment manufacturing facilities. Institutes could be involved in precompetitive stages.

The nucleus of an indigenous manufacturing equipment industry already exists in Korea. Companies such as LG have robotics and control business units. Korea's packaging assembly industry, its oldest electronics industry, already makes wafer handlers and positioners.

Whether or not the Korean government can spearhead future development such as the creation of an indigenous equipment supplier is uncertain. Little change is evident in funding philosophies of the various ministries. Rather than rolling back DRAM support in light of industry sentiment, the ministries appear to be augmenting these programs to keep annual spending constant. The justification for this is to keep university and institute research focused on key industrial problems in order to maintain the largest possible consultant and manpower pools.

Table 1.8. Highly Advanced National (HAN) Project

Category	R&D Project	Period	Technology to be Developed	Investment (US $ Mil.)	Project Planning and Coordinating Agency	Relevant Ministry
Product	1) New drugs and new agrochemicals	'92-'97	2-3 new antibiotics and germicidal agents	246	Science and Technology Policy Institute	MOST (MOHS)
Technology	2) B-Integrated Services and Digital Network (ISDN)	'92-2001	Core system and technology of B-ISDN	724	Korea Telecom	MOIC (MOST)
Development	3) HDTV	'90-'94	HDTV technology with movie quality level	12	Korea Academy of Industrial Technology	MOTE (MOST, MOC)
Project	4) Next generation vehicle technology	'92-2001	Electric car of 120km/h speed	563	Korea Automotive Technology Institute	MOTE (MOST)
Fundamental	5) Next-generation semiconductor	'93-'97	Basic & core technology related to ULSI memory chip	244	Science and Technology Policy Institute	MOST (MOTE)
Technology	6) Advanced material for information, electronics and energy	'92-2001	Generic advanced materials for information, electronics and energy	340	Science and Technology Policy Institute	MOST (MOTE)
Development	7) Advanced manufacturing system	'92-2001	Common and core technology for FMS, CIM & IMS	549	Korea Academy of Industrial Technology	MOTI (MOST)
Project	8) New functional biomaterials	'92-2001	High-quality and high-productivity biological resources	483	Science and Technology Policy Institute	MOST (MOAF)
	9) Environmental technology	'92-2001	Core environmental technology	289	National Institute of Environmental Research	MOE (MOST, MOTE)
	10) New energy technology	'92-2001	Fuel cell system and technology	357	R&D Management Center for Energy and Resources	MOTE (MOST)
	11) Next-generation nuclear reactor	'92-2001	Concept, basic design, and key technology	296	Korea Electric Power Corporation	MOTE (MOST)
Total	11 Projects			4,103		

1.4.6 International Cooperation and Recruitment of Foreign Nationals

Korea cooperates with other countries on a variety of levels, often with technology transfer being a stated purpose. Another deliberate method espoused by Korean government and industry to speed up technology transfer is recruitment of foreign nationals and overseas Koreans with knowledge and experience in high-technology fields. Korea has been actively recruiting U.S.-educated Koreans or persons of Korean descent and other foreign engineers to fill technology gaps.

Both MOST and KIST have made public their plans to hire foreign experts in order to raise technical standards and expedite development of the country's scientific and technological infrastructure. The Korean press has reported that the Korean government will match funds spent by companies recruiting and employing high-level foreign national personnel. These funds include salary incentives for qualified and accomplished foreign personnel at rates well above international standards. Reportedly, Korean government funding to help Korean industry recruit from the international "brainpool" was $3.4 billion in 1995.

Individual Korean companies have their own recruitment programs as well. Korean companies periodically take out full-page ads in the Silicon Valley newspapers and send recruiting teams around the world to scout talented engineers. Samsung Electronics recruited 150 PhDs in 1994, mostly high-level personnel already residing overseas. The company maintains a database on more than 1,000 overseas scientists and engineers.

At the same time as Korea's recruitment of foreign national experts has apparently intensified, the number of international cooperative programs has declined. From 1985-1990, Korea conducted 157 joint programs with Japan and the United States (31.4 programs/year); from 1990-1993 it conducted 101 programs (25.2 programs/year). More recent statistics were unavailable, but it appears that this reduction in activity has continued.

Of considerable interest is the dramatic increase in cooperative research projects with Russia, which amount to about 40% of the all international cooperative projects engaged in by Korean companies (Fig. 1.3). Cooperation with Russia may possibly be viewed as a way to augment Korea's internal resources with minimal threat to its economic security.

Middle managers interviewed in the course of this study indicated their companies' willingness to directly fund U.S. (and other countries') university centers, if these centers enhance production of trained Korean engineers. Collaborative research with leading universities ties in with Korea's global effort to open up and advance new technological frontiers. For example, the Electronics and Telecommunications Research Institute (ETRI) under MOST maintains relations with Columbia University, Stanford University, MIT, Carnegie Mellon, and the University of Illinois in the United States; University College, London and Oxford, in the United Kingdom; and HHI in Germany. Also underway are dialogues with several institutions in China, Japan, France, Australia, and Russia.

1985-1993

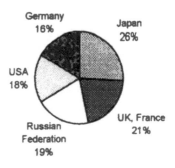

1993

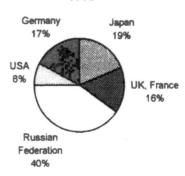

Figure 1.3. Korean joint research projects with other regions.

1.4.7 Government Assistance to Small and Emerging Businesses

In its institute structure, the Korean government attempts to assist small and emerging industries. MOTIE set up the Korea Electronics Technology Institute (KETI) in 1991 to assist in product development and to augment the tool base of companies struggling to get started. (See also Chapter 6 for more information about KETI.) KETI performs failure analysis and metrological services for relatively low fees. In product development, government will cost-share with companies through KETI; but the emphasis appears to be on more mainstream silicon-based technologies. KETI will serve the compound semiconductor community by forming links with external (most likely extranational) foundries. In this respect, this new institute appears to be in line with the development plans of the large, established Korean electronics firms, which are pressing to move out from the DRAM market into the ASIC field.

It is too soon to assess the impact of KETI or other government programs on small business development. In general, it appears that government policy is dominated by the development plans of the four major established "mainstream" chaebols. Emerging companies practicing non-standard technologies may suffer as a result. For example, the government has not assisted in the nucleation of a domestic GaAs capability, threatening the growth of small telecommunications companies like Dae Ryung.

1.4.8 Science Parks

In another means of indirect support of the electronics industry, the Korean government has played a major role in the development of "science parks." The most prominent of these is Taedok, located in Taejon. Construction of this government-planned city took place from 1973 to 1992. Taedok, located about 100 miles from Seoul, is populated by more than 10,000 scientists and engineers. Many research institutes and universities are located there, further increasing the concentration of R&D activities.

1.4.9 Industry Views on the Role of Government

Current industry sentiment runs against further government support of specific goal-oriented programs such as the DRAM and HAN projects mentioned above. While DRAM project support is relatively constant, industry cost sharing is declining significantly (about 30%, as projected from 1989-97). This is a result of a perceived maturation of the industry. As Table 1.9 shows, in the early 1980s period of DRAM development, most technology was purchased, initially from the United States and then from Japan. All internal development was precompetitive. At the time of this report, however, Samsung had become the world's highest-volume DRAM supplier, and LG and Hyundai were, respectively, in fifth and sixth places. Samsung announced a prototype 1 Gbit DRAM in December 1995 — ahead of any other company.

Semiconductor research is so highly competitive that no company is willing to put up with the reporting requirements of government programs, which amount to general disclosure, nor will companies agree to externally supplied schedules and milestones. Such considerations also impact industry-university relationships and industry's view of the role of government institutes. Korean industry is interested in research institutes primarily for augmenting university production of skilled engineers.

Table 1.9
History of Government Investments in ULSI (Up to 1 Mbit DRAM)

Development Period	1983-84	1984-85	1986-87
Method of Development	buying tech. from USA	buying tech. + development	DRAM

The model of technological development espoused by Korean electronics houses, as understood by this panel, holds that advanced technology does not transfer well, and therefore, R&D should ideally be undertaken internally. Even if adequate security safeguards exist, long transfer lead times and retooling difficulties outweigh the benefits of transfer. The relevance of the Collaborative R&D Promotion Law is called into question as a result. Of course, exceptions to the rule exist in other, less mature areas of Korean electronics, such as semiconductor tooling, equipment manufacture, and ASIC production.

1.5 THE ROLE OF THE UNIVERSITY

Korea's educational system is most impressive. It is extremely demanding and produces a very well educated population, as illustrated by Korea's literacy rate of 95%, one of the highest in the world. A high percentage of high school graduates enter universities, and roughly 20% of college graduates continue their education to obtain an advanced degree. *Business Week* reports that Korea has the highest number of PhDs per capita in the world.

It is an absolute national priority to educate the population at all levels, especially in math and sciences, so it is no wonder that Koreans have been able to make such strides in technology-based industries. Korean students pursue their academic goals enthusiastically, and they diligently shoulder heavy workloads throughout their six years of elementary school, three years of middle school, and three years of high school in preparation for college. University entrance examinations are extremely rigorous, as indicated by the term "admission war" to describe the fierce competition.

The Korean university system is a source of great national pride. The national commitment to education is shared by the government, the corporations, and the population at large. Koreans have developed their educational infrastructure to world-class standards.

Despite this accomplishment, the overwhelming majority of university faculty have earned their doctoral degrees at U.S. universities. Korean universities appear reluctant to recruit faculty from Korean programs, even if they consider their own graduates to be of comparable quality to those graduating from the West. This is a subject of considerable debate within the universities. Although they are a minority, excellent faculty who have graduated from Korean universities are teaching in Korea, and their numbers are growing. However, knowing the normal hiring preferences, many top students who intend to seek faculty positions in Korea prefer to get their advanced degrees in the United States. It is clear, however, that Koreans consider the quality of their education and facilities to be comparable to some of the best that the United States has to offer.

1.5.1 Facilities that Support Science and Technology

The Ministry of Education administers public universities, but the Ministry of Science and Technology (MOST) separately contributes funds to university science and technology (S&T) programs — both public and private — through the Korean Science and Engineering Foundation (KOSEF), Korea's equivalent of the U.S. National Science Foundation. KOSEF has established approximately 30 university S&T centers of excellence, which it funds annually at the million-dollar level. These centers are required to collaborate with at least three other institutions and are strongly encouraged to attract supplemental support from industry. [Swinbanks 1993] Among those centers with the highest reputations are the Korean Advanced Institute of Science and Technology (KAIST), Seoul National University (SNU), and the corporate-sponsored Pohang Institute of Science and Technology (POSTECH). (See Chapter 6 for more details on these three centers.) KOSEF provides the centers with support for a broad range of scientific disciplines, including physics and engineering. Some of the S&T centers broadly support semiconductor materials processing. SNU, for example, has a Research Center for Thin Film Fabrication and Crystal Growing of Advanced Materials, and KAIST has a Material Surface Engineering Center [Swinbanks 1993].

KAIST is unique in that it is the only national university administered under the auspices of MOST. This has allowed KAIST to skirt many government rules related to hiring that have been considered an impediment to recruiting top faculty at other public institutions. For example, KAIST is able to offer triple the salary of other universities to recruit the best professors worldwide. Also, the curriculum emphasizes research in applied fields that are defined by MOST as national priorities [Swinbanks 1993].

Other public universities, such as SNU, have excellent facilities and highly regarded faculty. Because historically SNU has been the premier Korean university, it continues to attract the best student population, and although it can not compete with KAIST's financial incentives, it boasts one of the most prestigious faculties in the country. SNU operates a first-rate facility for teaching semiconductor processing. It is a 4-inch facility that can run a full CMOS line with 1.5 micron design rules. The equipment there is comparable to that of the University of California at Berkeley, of the Massachusetts Institute of Technology, or Stanford. The facility even has an e-beam direct-write lithography system that is regularly used in experimental fabrication runs. In fact, it routinely processes runs of multiple project chips with designs from other universities. The fact that SNU's facility regularly puts through working devices, starting with bare silicon and ending with packaged parts, puts it at the top of the world in teaching fabrication lines.

Also offering excellent S&T education in Korea are the privately funded universities established by industry. Pohang Steel Company's university, POSTECH, Daewoo's Institute of Advanced Engineering (IAE), and Samsung Advanced Institute of Technology (SAIT) are attracting excellent faculty and highly qualified Korean students [Swinbanks 1993]. An unstated goal of the company-owned schools appears to be to build loyalty with the top engineering students at the same time as training them, so that they will come back and work for the company.

Of the 150 colleges and universities in Korea, approximately 100 have electrical engineering departments; 70 of those are active in the Integrated Circuit Design Center (IDEC). Of the institutions active in IDEC, 40 are well regarded by the Korean semiconductor industry for their teaching of IC design. Graduates from all these institutions can count on finding high-quality jobs in their field in Korea.

The availability of cleanroom facilities at Korean institutions of higher learning, programs like IDEC, and the excellent teaching programs at KAIST, SNU, and other fine technical universities, puts Korea on par with United States for educational resources to support the semiconductor industry. Electrical engineering students emerge from Korean programs with strong backgrounds in solid state device processing, design, and layout. While industry does emphasize support for applied research at universities, this does not suggest a weakness in the basic sciences: through KOSEF, the government primarily supports research in basic science.

1.5.2 Industrial Connections

Korean industry takes the position that a well-educated workforce is critical to competitiveness. Corporations generally sponsor research of considerable educational value, even if the practical applicability may not be immediately apparent. Industry leaders do not believe that research of individual university professors working with small groups of graduate students can lead to commercially important innovations. They believe that innovation is the product of the perseverance and dedicated work of many highly skilled researchers striving as a team toward a common goal. Nonetheless, they do strongly believe that university research produces well-qualified graduates and therefore well-qualified employees — that is, highly skilled and innovative engineers.

Approximately half of the research money in the most highly regarded electronics programs such as those of KAIST and SNU comes from industry. The universities then engage in the balancing act of trying to accommodate the research needs of professors while trying to train students that meet the needs of industry. (For students who are sponsored directly by a company, there is an understanding that that company will have the first chance to hire them when they graduate.) Company executives develop close relationships with individual professors in order to have confidence in their abilities to attract and train top students that the companies will ultimately employ. As a consequence, professors often have a great deal of clout within the industry.

As the success and profitability of large Korean companies have increased, several have started their own educational institutions, some of which are full-fledged universities that offer degrees up to the PhD level. There appear to be at least three factors contributing to this trend of companies founding universities: dissatisfaction with the political climate within public Korean universities; a desire to better guide the training and recruitment of a skilled workforce; and the philosophy of diversifying whenever and wherever possible. Private industry's universities are completely independent of government agencies and funded solely by the parent companies. The model appears to be similar to the U.S. universities started in the early industrial revolution by industrialists such as Andrew Carnegie and Peter Cooper to support their own rapidly growing industries.

One example of a company starting a world-class university from its own profits is Pohang Steel Company, which owns and operates POSTECH. POSTECH, as do other private schools, uses financial incentives to attract the best students. In less than a decade since POSTECH opened its doors, it has come to attract the top 2% of the nation's college students [Pettit 1989]. Also like other industry schools, POSTECH pays competitive salaries to its faculty, much higher than those offered at government-run schools and even more than at KAIST. POSTECH boasts having mostly U.S.-trained faculty, graduates of such schools as UC Berkeley, Stanford, and MIT.

Although Korean companies enjoy the prestige and influence over curriculum afforded by running their own universities, they do see a distinct role for the government. Government can better focus the research it sponsors to meet national priorities. The government is also better able to sponsor university research geared to developing more future-oriented, precompetitive technologies. Some companies argue that government is too heavily involved in commercially competitive technologies and that more of its resources should be applied to basic science.

1.5.3 Educational Goals of University S&T Programs

An overwhelming proportion (>80%) of faculty at the more prestigious Korean universities are educated in the United States. Many U.S.-trained faculty members in Korea appear to be committed to teaching global awareness and creative, independent thinking skills as well as those technical skills their students will need. Faculty generally agree that such new thinking skills are necessary if the country is to truly move ahead in a high-tech world. They also realize that they have a difficult job to overcome cultural hurdles in order to nurture a more creative workforce for the future. The Ministry of Education also realizes this and has begun implementing changes to foster creative and independent thinking skills at the secondary and university levels, again with the goal of promoting self-reliance in high-technology skills.

Some Korean faculty express concern that their graduates are unprepared for the demands on creativity required to make new designs and more competitive products, anticipating that "the future does not remain with DRAM, but with sophisticated and advanced circuit design." They acknowledge that the requirement to be competitive in the future is fostering novel ideas that combine all fields related to electronics manufacture (layout, architecture, systems, fabrication, analog, and others), and that it will take imaginative and resourceful minds and team efforts to achieve interdisciplinary solutions. This perception of vulnerability is contrary to the authors' impression that the Korean workforce is very well prepared to work in today's competitive semiconductor industry. Perhaps this perception helps to keep Koreans forging forward with such diligence.

Professor Chong-Min Kyung, the newly appointed chairman of the electrical engineering department at KAIST, observes that "Korean culture is responsible for the lack of teamwork and long-term planning, two deficits of the educational system." (This statement comes from a man who is determined to make his department one of the best in the world.) To address such concerns, KAIST intends to encourage better collaboration by building large projects with 3 to 5 professors from different disciplines and their students, who will all work together to complement each others' expertise. The projects will generally be practical in nature in order to satisfy the sponsors, who generally do not have long-term goals in mind; nonetheless, the departments will do their best to encourage aspects of the research that explore more fundamental insights and develop the sense of unforeseeable goals.

As developing an equipment manufacturing infrastructure becomes a higher national priority, some combination of industrial and university research will be ready to take on the challenge. This area is one in which the government may take a stronger role, for it can sponsor university research within many disciplines related to semiconductor manufacture without conflict from the universities; the results would benefit all concerned.

1.6 INDUSTRY/GOVERNMENT/UNIVERSITY INTERACTIONS

In general, representatives from the industry, government, and academia agree that their respective organizations have interacted very well to advance the global standing of the Korean semiconductor industry. However, assessments of each organization's accomplishments do differ. Some industry leaders believe that government and academic R&D have not directly helped the industry. They would like the universities to focus on supplying trained manpower, and the government and academic R&D labs to focus on performing more risky, long-term research. On the other hand, government and academic leaders believe that they have made significant contributions to the semiconductor business in Korea.

The government organizes and funds focused projects in which industry, R&D labs, and universities collaborate. This approach has led to successes such as the launching of a Korean satellite and on-going building of an advanced telecommunications network in the country. All the major semiconductor companies participate in the government's long-term program on semiconductor memory R&D, although it is not clear to what extent each company participates. However, companies are increasingly relying on their own R&D as the technology becomes advanced and product orientation becomes the most important issue in R&D. Overall, it appears that the government planning and coordinating roles have been beneficial to the electronics industry by minimizing redundancy of effort and synchronizing development in related areas.

Some electronics industry executives believe that the government has contributed indirectly to the growth of the industry by making R&D less expensive through tax write-offs. However, there are those in industry who believe that the Korean government could do more to help Korean companies on the economic policy issues, that issues such as granting licenses for new manufacturing plants or raising capital by limited stock offerings are sometimes resolved more on the basis of political pressures than on technical or economic merits, and that bureaucratic inertia (etc.) slows decision-making processes that involve industry.

The universities, colleges and research institutions funded by the government were considered by company representatives to be doing an adequate job of providing a well-trained workforce for the semiconductor industry. All companies provide scholarships at the universities and send employees for training at the universities for periods extending to 6 months or more. Companies provide funds to universities either directly or through government-sponsored programs. For example, a donation from Hyundai fully paid for a new, well-equipped CAD training center at the Korea Advanced Institute of Science and Technology (KAIST) that will be open to students and visitors from the industry.

Despite the complexity of industry-government-university inter-relationships in Korea, achieving a level of interaction that is advantageous to all appears inevitable for several reasons: first, a lot of thought seems to be given to this topic; second, Koreans can develop models based on lessons learned from the United States and Japan; third, cohesiveness of the culture and national goals is a powerful catalyst; and finally, these entities are mutually interdependent. Korea's semiconductor industry cannot maintain its high growth rate without the supply of highly trained professionals from the institutions, and it cannot globalize without the government's assistance in international affairs. Likewise, the success of the semiconductor companies is essential to the success of the institutions and the government.

1.7 SUMMARY

Korea has found the use of technology transfer methods and the integrated nature of its globalization strategy to be successful in its preliminary competitive development stages, enabling it to advance its technological infrastructure more rapidly than previously thought possible. The country now has in place a systematic, integrated globalization strategy for rapid economic development that harnesses the strengths of its private sector, academia, and government. An example is the improvement of assimilated foreign technology, specifically the 4Mbit DRAM technology transferred to Korea in the early 1980s. The chaebols improved and built upon the original chip, creating a faster, higher-memory 16 Mbit DRAM chip that was then mass-marketed to the United States, and they continued to evolve their technical competence to the point that they were able to develop the world's first 1 Gbit DRAM ahead of all competition. Thus, Korea has proven its technology transfer and production expertise, but also its incipient innovation capabilities.

Before all of the benefits of technology transfer and globalization can be realized, Korea's business and political leaders believe the national R&D strategy must fully transition from a "catch-up" to a "creation" strategy, where innovation is supported and nurtured. Current sentiment indicates that this transition has been made and technology transfer (primarily from abroad) will no longer be a major issue in the development of the electronics industry. One possible exception to this is in the area of equipment innovation and development. It is interesting to note the large increase in cooperative programs with the former Soviet Union. This is viewed as an attempt to augment a busy work force.

Korean government, industry, and academia are all working together to support the goal of the nation becoming a technology innovator of global stature, and rapid progress is being made. Still, some significant infrastructural obstacles remain, not the least of which is the economic dominance of the huge chaebol conglomerates that has tended to restrict the viability of innovative small- and medium-sized enterprises.

Chapter 2

THE KOREAN SEMICONDUCTOR INDUSTRY

Memory technology dominates the Korean electronics industry. Looking into the memory industry's products, production, sales, and current research and development efforts gives insights into the present strengths and weaknesses and the possibilities for the future of Korea's semiconductor industry.

2.1. KOREA'S MEMORY INDUSTRY

Korean semiconductor manufacturers have a very small domestic market compared to their international one, and international markets tend not to be as forgiving as domestic ones. Therefore, it is especially important that manufacturers choose their investments in technology wisely and tool up for large-scale production of products that will sell well in a global marketplace. For these reasons it is important to look at how Koreans perceive global trends and their place in them.

Market statistics show that in 1995 the world semiconductor market was worth $150 billion, up 40% from the previous year. Numerous analyses are optimistic concerning the continuing growth of this global industry as the use of computers and related devices proliferate, as the memory requirements of these devices increase, and as the technology moves to new applications. This optimism is demonstrated by Table 2.1, which shows the growth and projected growth on a global scale of various memory technologies.

Within this framework of global growth, the Korean domestic market itself has been experiencing impressive growth. Figures provided by the Korean Semiconductor Industry Association (KSIA) show that total sales in the domestic memory industry in 1995 were $15.8 billion, up 89% from the previous year. (It is important to note, however, that 1996 sales and income from 16 Mbit DRAMs, in particular, have been way down due to world market conditions.)

Table 2.1.
Sales and Projected Sales for Global MOS Memory Market
(in $ millions)

	1993	1994	1995	1996*	1997	1998
DRAM	14,411	19,654	17,767	18,028	20,839	23,619
SRAM	3,900	4,140	4,590	5,500	6,250	6,770
EPROM	1,360	1,150	925	850	745	635
FLASH	590	1,150	1,925	2,850	3,650	4,430
MASK ROM	1,835	1,985	2,075	2,480	2,670	2,845
EEPROM	390	425	450	475	500	500

* As of October 1996, it appeared that these figures would be lower than anticipated.
Source: KETI 1994

Meanwhile, as a memory producer, Samsung jumped from third place in 1992 to first in 1993, making Samsung the first non-Japanese company to rank number one in the memory market since 1981. Table 2.2 shows the sales for all Korean manufacturers from 1990-1995, and also indicates the percentages of sales that are in the domestic and overseas markets. This demonstrates how sensitive the industry's revenues are to the global market, and how imperative it is for Korea to excel in the technologies that shape market trends.

Table 2.2
South Korean Semiconductor Sales (in $ millions)

		1991	1992	1993					1994	
				1/4Q	2/4Q	3/4Q	4/4Q	SUM	1/4Q	Growth rate, %
DRAM	Production	1,506	2,182	731	881	1,013	1,159	3,784	1,267	173
	Export	1,367	2,067	683	829	948	1,089	3,549	1,168	171
SRAM	Production	163	294	80	107	118	118	423	130	164
	Export	145	267	71	96	106	107	380	119	169
Other	Production	127	167	42	51	67	90	250	93	221
	Export	108	140	34	43	60	80	217	80	235
Sum	Production	1,796	2,643	853	1,039	1,198	1,367	4,457	1,490	175
	Export	1,620	2,474	788	968	1,114	1,276	4,146	1,367	174

* Assembly excluded
Source: Semiconductor Industry, May 1994

In terms of expansion, by mid-1994, Korean manufacturers had invested considerable capital into putting on-line mass production fabrication equipment capable of 0.5 µm resolution on 8-inch silicon wafers. Samsung has one 6,000 wafer/month and one 14,000 wafer/month system under its No. 5 assembly line. Hyundai has put into operation a Fab. 4 and a Fab. 5 line, each with a capacity of 10,000 wafers/month. LG Semicon has a 10,000 wafer/month capacity with its C2 assembly line. Table 2.3 summarizes each company's new production capacity.

In an effort to develop technologies in the domestic equipment and materials industry, plans are underway to attract U.S. and Japanese manufacturers to the Chon-an area of the Republic of Korea. As part of this endeavor, Applied Materials Korea Company (100% U.S.-owned) set up the Technology Center in 1995, and DNS Korea was started as a joint venture between Samsung and DNS (the Great Japanese Screen Company). In another joint venture arrangement, Anam announced in September 1996 a $3 billion contract with Texas Instruments (TI) to conduct fabrication and packaging processes for nonmemory semiconductors, with TI transferring to Anam 0.35-micron processing technology and CMOS technology, and with joint plans to commercialize 0.25-micron, 0.18-micron, and 0.13-micron processing technologies in coming years.

2.2. DRAM

South Korean electronics industry leaders foresee a need for a faster, lower-power, and more heavily integrated DRAM in order to improve the performance of computers in the PC and workstation markets. They see the use of R-DRAM (Rambus DRAM), S-DRAM (Synchronous DRAM) and C-DRAM (Cache DRAM) as important for bringing memory speed closer to microprocessor speed.

In terms of DRAM improvements, industry leaders see more efficient usage of memory cell real estate as an important way to increase capacity — specifically, through a combination of stack and trench type cell structures. Moreover, they see enhancement of transistor and capacitor fabrication technologies as being important to further improvements.

Table 2.3
South Korean Manufacturers' Processing Capability
for 0.5 µm 8-inch Wafers in 1993

	Samsung		Hyundai		LG Semicon	
	Line	Monthly Capacity	Line	Monthly Capacity	Line	Monthly Capacity
	1	6,000	FAB.4	10,000	C2	10,000
	2	14,000	FAB.5	30,000		
SUM		20,000		40,000		10,000

Source: Il-kyoung Micronics, September 1993, and HEI

This necessitates designing chips that operate at lower voltages in order to ensure their stability as transistors and capacitors become smaller (the large portable PC market also contributes incentives for developing low-powered memory). In addition, wide I/O structures are seen as the future standard, since this shift can lower power dissipation, decrease chip sizes, and reduce costs. Also, the industry tends to see the use of DRAM in telecommunications and video equipment as a force that will tend to move manufacturing away from mass production and more towards customized production.

Despite these trends in DRAM improvements, the Korean semiconductor industry is experiencing a gradual loss of efficiency in terms of capital investment versus returns for the production of DRAM. In the long run, industry leaders anticipate the eventual replacement of DRAM with flash memory in a variety of applications.

Table 2.4 gives the status of the 1993 DRAM market, which shows that Samsung experienced a remarkable growth of 72% and that Korean manufacturers make up almost 30% of the revenues earned by the top 10 DRAM producers in the world. It must be noted that as the global DRAM market experienced a glut in 1996, it is unclear what kind of position the Korean manufacturers have in the market at present, but general trends are likely to hold true. Korean electronics manufacturers are proceeding with plans to move into the higher-performance, higher-value-added markets.

The strengths of Korea's firms in this area include their capacity to produce high-speed, highly integrated (such as 64 Mbit) memory packages.

Table 2.4
Top 10 Companies in the DRAM Market, Based on 1993 Performance
(in $ millions)

1993 Rank	1992 Rank	Company	1992 Revenue	1993 Revenue	Growth Rate(%)	1993 Share (%)
1	1	Samsung	1,192	2,046	72	14.0
2	4	Hitachi	824	1,567	90	10.8
3	3	NEC	894	1,520	70	10.4
4	2	Toshiba	1,123	1,479	32	10.1
5	N/A	IBM	N/A	1,133	N/A	7.8
6	5	Texas Instruments	667	996	49	6.8
7	6	Mitsubishi	628	930	48	6.4
8	10	Micron Technology	445	793	78	5.4
9	9	Hyundai	448	706	58	4.8
10	8	LG	513	689	34	4.7

Source: KETI 1994

Also, they are studying the use of Ta_2O_5 as a material to be used as a thinner dielectric and insulator (also to have the effect of increasing capacity). In addition, they are looking to R-DRAM, S-DRAM, and C-DRAM to bring DRAM speeds on a par with microprocessor speeds.

The weaknesses of Korea's memory industry include its slowness relative to competitors to develop higher-pin-count and lower-voltage products. This slowness in turn hinders efforts to develop application-specific integrated circuits (ASICs) and memory packages with special functions.

2.3. SRAM

Development requirements for electronic static random access memory (SRAM) technology closely parallel those for DRAM: faster speeds, larger capacities, lower power dissipation, and wide I/O architectures. Since it doesn't require periodic refreshment, SRAM is perceived by Korean engineers to be more appropriate than DRAM for applications involving lower power dissipation and increased speed.

More specifically, in terms of market trends, industry leaders believe a 10 ns or shorter access time is needed and achievable with SRAM. BiCMOS, which can reduce access times by 30-40% over existing CMOS, is seen as a promising means of increasing speed. To improve capacity, more heavily integrated architecture and pseudostatic DRAM technology are considered to be promising. Lead-on-chip architecture is also seen as a key method of reducing chip size and increasing capacity. The industry looks to the wider I/O market to soon supplant the X1/4 I/O standard. Finally, a move from resistance-type to thin-film-transfer (TFT) type structures is seen as the most promising means of reducing voltage requirements.

Table 2.5 ranks the top 10 SRAM manufacturers. Samsung, with sales of $200 million in 1993, is the only Korean company to make it to this list. The strength of Korean semiconductor companies in this market is their ability to produce highly integrated products. Korean firms are six months to one year behind market leaders in producing BiCMOS, high-speed CMOS, high-pin-count, and special-function SRAMs; however, they do seem to be making improvements in basic R&D.

2.4. EPROM, EEPROM, AND FLASH MEMORY

South Korean companies are absent from all three programmable non-volatile memory technologies; no Korean manufacturer has a major stake in either EPROM or erasable electrically programmable read-only memory (EEPROM). The special equipment required for reprogramming EPROM and the very high bit-cost of EEPROM make these markets unappealing.

Table 2.5
Top 10 Companies in SRAM Market
(in $ millions)

Rank	1993 (Estimated Figures)	
	Company	Sales
1	Hitachi	575
2	Toshiba	300
3	NEC	290
4	Fujitsu	250
5	Motorola	205
6	Samsung	200
7	Sony	195
8	Mitsubishi	170
9	IDT	155
10	Sharp	135
10	Cypress	135

Source: KETI 1994

Korean industry is, however, putting a premium on becoming a major player in flash memory, and this design is a major factor in the strategy of Korean firms to lay claims to future memory markets, since flash memory is expected to be a key market after 1996. Table 2.6 lists the advantages and disadvantages of DRAM, SRAM, and flash memories.

Table 2.6
Advantages and Disadvantages of Major Memories

	Advantages	Disadvantages	Applications
DRAM	• low bit cost • small size, more pins • general use	• limits in speed • required surrounding circuits • volatile	• PC, WS • main memory • OA machines
SRAM	• high speed • low power dissipation • simple circuits	• high bit cost • large chip size • volatile	• high-speed memory • main memory for super computers • portable OA machines
FLASH	• adequate for integration • non-volatile	• limits in rewriting (currently ~100,000 rewrites) • low speed	• IC cards • alternative fomagnetic media

Source: NRI

Some in the industry foresee flash memory as the eventual successor to DRAM in diverse applications, for a number of reasons: Flash memory is lean on power consumption, and the cells are much smaller and do not require separate capacitors, thus allowing for very high capacities. All this makes flash memory attractive for a variety of computing needs, especially ones involving portable machines. In addition, flash memory is attractive in terms of requiring a relatively small facility investment, while allowing for highly integrated, low-cost mass production. Based on the promising characteristics of flash memory, Samsung, as an example, has broad expectations for the memory's uses, as Table 2.7 shows. With their development of NAND-type flash memories, the Koreans hope to grow in the flash memory market.

In addition to pursuing development of flash memory, the Korean memory industry also appears interested in ferroelectric random access memory (FRAM). The technology is touted as having the advantages of both DRAM and flash memory, due to being nonvolatile like flash memory, but allowing the freer bit-access and bit-rewriting capabilities of a RAM. The technology does have a number of major difficulties, which make some in the global industry skeptical; however, many in the Korean industry believe these problems can eventually be overcome to give FRAM a significant market share in the next 10-15 years.

2.5. MASK ROM

The emphasis on flash memory that Korean semiconductor companies appears to be making does not mean they are deemphasizing the less exceptional mask ROM (read-only memory) technology. Indeed, with mask ROM's advantages of high capacity at a reasonable price, it is expected to continue to be a central member of the memory market and a popular constituent of consumer applications such as game machines, karaoke machines, electronic note pads, and electronic dictionaries.

Table 2.7.
Uses of Flash Memory

Period	Existing System	Alternative System
Short Term (1993-1997)	BIOS (basic input/output system) Locative Structure of EPROM Locating Disk in DOS	Flash Memory (Component)
Mid Term (1997-2000)	Floppy Disk & Hard Disk Driver	Flash Card Solid State Disk
Long Term (2000-)	Main Memory (DRAM)	Flash Memory (Module/MCM)

Source: Samsung Electronics Co.

In the future, mask ROM is expected to compete with high-density products like CD-ROMs in such applications as fax machines and printers.

Table 2.8 lists the top five mask ROM producers in the world, which includes one Korean manufacturer, Samsung. The mask ROM market is dominated by Japanese companies such as Sharp and NEC, whose combined share was about 50% in 1993. Between 1992 and 1993, Samsung moved from 5th to 4th place with a 60.7% increase in revenues.

Korean firms are competitive in the production of high-capacity and even special-function mask ROM memory modules; however, they lag the market by about half a year in the production of high-pin-count products.

2.6. ADDITIONAL SEMICONDUCTOR TECHNOLOGIES

The relatively small semiconductor industry outside of memory technology focuses mainly on areas such as digital signal processing and telecommunications. For example, VLSI design programs have focused on low-powered integrated circuits for vocoders, digital cellular communications, and asynchronous transfer mode physical layer processing. Meanwhile, projects extending into ASICs and compound semiconductors have been enacted, due to an expected strong market in these technologies in the communications field. Much of the research is conducted within the Electronics and Telecommunications Research Institute (ETRI). The work includes the following:

- For variable-rate vocoder applications, a 16-bit DSP processor and vocoder software have been developed. The processor's instruction set and internal architecture were optimized for vocoder algorithms.

Table 2.8
Sales and Market Share of Top 5 Companies in Mask ROM Market

Rank	1992			1993		
	Company	Sales ($M)	Market Share (%)	Company	Sales ($M)	Market Share (%)
1	Sharp	390	28	Sharp	475	26
2	NEC	330	23	NEC	400	22
3	Hitachi	190	13	Hitachi	230	13
4	Fujitsu	145	10	Samsung	225	12
5	Samsung	140	10	Fujitsu	170	9
Others		215	15		335	18
SUM		1,410	100		1,835	100

Source: KETI 1994

- In digital cellular communications, a CDMA (code-division multiple access) modem chip set has been designed that consists of an ASIC for use in CDMA mobile terminals and three ASICs for CDMA base stations. Extensive modeling and simulation of algorithms and hardware design were carried out for functional verification, and a clever circuit design has helped reduce chip area and power consumption.

- For digital video signal processing, a video signal processor architecture and a video ADC/DAC (analog-to-digital/digital-to-analog converter) have been developed. The video ADC and DAC were implemented with 5 V CMOS technology. The video ADC has a 10-bit resolution at a 20 Mbps rate and consumes less than 120 mW at 20 Mbps and 100 mW at 15 Mbps. The video DAC has a 10-bit resolution and operates at 80 MHz.

- For ATM physical layer processing, a single chip of 155.52 Mbps and a three-chip set of 622.08 Mbps have been designed. These chips were implemented with 0.8 mm CMOS technology.

In addition to the above, the Korean semiconductor industry foresees a growing need for high-speed and large-capacity telecommunications systems. To meet this need, use of compound semiconductor devices is believed to be inevitable, and they are expected to play an important role in the coming information era because of their inherently superior high-speed electronic and optical properties compared to silicon devices. To cope with such needs, ETRI's Compound Semiconductor Department has carried out numerous research projects to develop electronic and optical III-V devices.

2.7. PROCESSING TECHNOLOGY

For the near future, the Korea still looks to photolithographic techniques for mass production — mainly to lithographic design improvements such as shortening the wavelength of the stepper, increasing the numerical aperture, and improving the quality of the resist. However, many in the industry also see that for the massively integrated circuits (e.g., 1 Gbit DRAM) of the post-2000 era, current fabrication technologies will be inadequate, thus requiring development of X-ray or other next-generation lithographic techniques.

In etching technology, the industry emphasizes plasma etching for mass production and will improve techniques in this area as a means of improving precision. To improve the ion infusion process, researchers are working to improve the homogeneity of the wafer surface, especially by using the parallel beam scanning technique. In metalization technologies, researchers perceive the next-generation materials that will replace aluminum are copper, gold, and tungsten. With regards to thin film formation technology, the CVD process is expected to become a widely used alternative to the sputter method for Al wiring.

In terms of packaging, the Korean industry is focused on smaller, lighter. and more sophisticated packaging, with the expectation that current technology will soon prove inadequate. Table 2.9 sums up the global trends in packaging issues as seen by the Korean industry.

In particular, Korean industry leaders see a growing demand for very thin packaging for memory cards to use with portable machines such as notebook PCs, video cameras, and cellular phones. Packaging technologies being developed to meet this demand include lead-on-chip (LOC), 3-D packages, and chip-sized packages.

2.8. SHORTCOMINGS AND SOLUTIONS

Korean electronics manufacturers perceive that a major problem they face is that they lag their competitors in domestic development of basic manufacturing and materials technology. As a result, they must import much of their manufacturing equipment and materials technology from oversees. Korean industry analysts view Korea's electronics technology profile as being rather top-heavy, having a competitive edge only in some high-technology areas. Korea's basic domestic technological standards are deemed to be far behind those of their major East Asian competitors, Japan and Taiwan. In addition, except for bonding techniques and testing software, Korea's fabrication processes are still behind the standards of their competitors.

Company executives believe that these shortcomings have had some deleterious effects on their companies' competitiveness. For example, due

Table 2.9
Global Packaging Trends

Devices Trends	Packaging Trends	Technological Issues
High capacity Sophisticated functions	Larger chips high pin count	New packaging technology: LOC, flip-chip, chip-size packages, 3-D memory, high-precision lead processing
Minimization	Smaller packaging Thinner packaging	Lowering transformation of resin High precision of mold and processed mold
Customized production	Diversification	Automated processing and customized production

Source: *The Status and Prospects of the Semiconductor Memory Industry*

to weaknesses in manufacturing design techniques and state-of-the-art precision equipment technologies, domestic manufacturers were unable to respond quickly to the emerging trends for higher-pin-count DRAMs, and as a result they consistently trail market leaders in introducing new wider-I/O products. In addition, in the area of materials technology, they lament that their grasp of metal wiring and leveling techniques is significantly behind many of their competitors, again hindering their production of more sophisticated products.

2.9. STRENGTHS

The most important strength of Korean semiconductor companies seems to be their employees. Company brochures and propaganda videos point out the importance of employees. Employees create value for the customers. An army of enthusiastic, well-trained professionals, some of whom work tirelessly for 72 hour each week, is a national asset.

Excellence in manufacturing technology with rigorous implementation of TQM (total quality management) and statistical process control principles and ISO (International Standards Organization) certification is another important factor that strengthens Korean companies. Ability to implement strict manufacturing discipline in a large group of motivated employees plays a major role in achieving high yields. Manufacturing excellence ultimately leads to low costs per unit and increased sales.

Good earnings and willingness to invest heavily in R&D also provide strength to Korean companies. Korean R&D expenditure increased from U.S. $2 billion in 1990 to about U.S. $8 billion in 1995.[Asia business news, Korean cable TV, Dec. 14, 1995] This trend is expected to continue for product-oriented R&D. Although some funds are invested in catching up with world technology leaders, overall, this R&D commitment is handsomely paying off for Korean companies and assuring global technological leads. An MPEG chip made and marketed by LG is considered a development of worldwide importance. According to an LG spokesperson, U.S. companies now want to license LG's MPEG technology. Samsung and now LG Semicon have announced technologies for 1 Gbit DRAMs well ahead of the competition.

R&D investment by Korean companies has a global component, as evident in figures published by the Washington DC office of the Korea Foreign Trade Association: Republic of Korea enterprises operated 27 R&D centers in the United States in 1993 — 10 in semiconductors, 7 in computers, 4 in high-definition television, 3 in automobiles, and 1 each in software, telecommunications, and biotechnology. According to this 1993 data, Korea ranked seventh in the list of countries with research centers in the United States: Japan (225 centers), Britain (109), Germany (95), France (52), Switzerland (47), Netherlands (29), and Korea (27).

Some economic issues strengthen Korean companies' position in the international marketplace. The relatively low value of the Korean *won* in comparison to the U.S. dollar and the Japanese yen has helped keep the cost of Korean goods low in the international markets and thus contributed to increased sales, as do lower labor costs, although the authors did not review salaries or benefit packages in Korea.

The semiconductor companies appear to derive considerable strength from Korean culture and traditions, a fact invisible to the first-time visitor. Loyalty, pride, hard work, striving for technical excellence, and the resolve to be number one all contribute to the high productivity that improves the rankings of the companies year after year. Projection of a giant family image and its reinforcement with broad employee benefits such as housing, dormitories, and scholarships, with a uniform dress code, which portrays equality, and a promise of a life-long company-employee relationship nurture employee loyalty and increase productivity.

Engineers and middle level managers point out that strong leadership and good capital support from parent companies has played an important role in the success of Korean semiconductor companies. The persistence expected of engineers in the trenches is also exhibited by company leaders. They take risks, make long-term investments, and maintain stable, clear, and consistent goals for middle management. Corporate culture and the management philosophy discussed below play a strong role in keeping the workforce motivated and thus contribute to the company's success.

The exacting Korean work ethic, combined with other strengths discussed above, have provided all the horsepower Korean electronics corporations have needed. Right people when provided with right training, right tools, right resources, proven technology, strong leadership with perseverance, and shared goals have allowed Korean companies to surpass performance expectations year after year.

2.10. STRATEGY FOR COMPETITIVENESS

In addition to sharing the national ambition for Korea to become an international economic power house, each of the four major Korean semiconductor companies has the ambition of becoming one of the top five semiconductor companies in the world by the next century. Based on 1994 data, the Korean giants already rank very high in DRAM production in the world: Samsung is number one, Hyundai is number two, and LG is number six. They are keenly aware of the competition among themselves for market share, and of the competition with other Asian countries such as Taiwan, which is trying very hard to get a significant share of the DRAM and ASIC markets. Engineers from all four major companies feel that Taiwan is on its way to becoming a major semiconductor component supplier to the world. In fact, they are concerned that Taiwan could hurt the Korean semiconductor industry as Korea hurt the Japanese semiconductor industry and as Japan hurt the U.S. semiconductor industry. But the focus and concern in each Korean firm's strategy seems to be global competitiveness and dominance in the broader international markets for electronics goods.

Koreans also know that they need to leapfrog in technology to compete with innovative U.S. companies. The Korean Semiconductor Industry Association (KSIA), Electronics Industries' Association (KEIA), Korean academic institutions, the Korean Ministries MOST, MOIT, and MOTIE, and the companies themselves have all examined the competitiveness of the Korean semiconductor industry. Although a blueprint for a national strategy has not emerged, the discussion below is based on themes apparently held in common by the various organizations, as observed by and discussed with the panel authors.

Know and leverage your strengths. Koreans are proud of their manufacturing capabilities; their pride is reflected in the low defect densities emphasized by Signetics-Korea, in LG Semicon's claims to have the best DRAM manufacturing yields in the world, and in Hyundai's Corporate Product Display Gallery, which exhibits numerous Hyundai products based on foreign technology. Koreans invest heavily in maintaining their manufacturing excellence. Automating, upgrading, and building new facilities; dedication to total quality management; extensive training; hiring and/or repatriation of ethnic Korean foreign professionals; and encouragement of workers to maintain the Korean work ethic all strengthen manufacturing excellence. Manufacturing excellence is supplemented by non-union, non-Western management philosophies and company-wide motivational campaigns. Korean managers appear to believe that the companies' manufacturing strength results from the contributions of ALL employees, and that the companies leverage this strength by providing job satisfaction, ample benefits, and by evolving corporate maxims on the principle of valuing employees.

Korean companies are getting ready to make large profits in the exploding semiconductor market by spreading their manufacturing excellence worldwide. LG Semicon, in collaboration with Hitachi, is building a fabrication facility in Malaysia; Hyundai is building the world's largest fabrication facility in the United States (in Oregon); Daewoo is building a fabrication facility in Europe; Samsung is building a facility in Texas; Samsung and Hitachi are collaborating on building a new facility in Asia (location not yet announced).

Know your weaknesses and eradicate them. Koreans in this field appear to be keenly aware of the lack of technology infrastructure, design skills, system integration expertise, state-of-the-art R&D, and technical innovation. They acknowledge dependence on foreign technology and the transient nature of the wealth brought in by the volatile DRAM market. Hence in every organization (universities, research labs, government ministries, KSIA, and the companies themselves) there is a strong commitment to establishment of a semiconductor industry infrastructure. This commitment is manifest in new R&D centers, foreign collaborations, technical training at U.S. universities, and hiring and repatriation of U.S.-trained professionals. Two academic IC prototyping centers are now operational. The four semiconductor giants have agreed to serve as foundries for fabrication of chips designed by R&D center researchers and trainees.

Learn from others. It is often said that in semiconductors the Koreans are following the example of the Japanese. Koreans go beyond this view and explain that they would like to avoid the mistakes made by the Japanese. They would not let the manufacturing leadership slip to another nation such as Taiwan. They would guard against trade friction with the United States by building manufacturing plants in the United States. They would avoid over-capacity in component manufacturing by becoming less dependent on the component market and moving into the systems market. Koreans are literally making lists of technology fields to dominate, and they are emphasizing the establishment of infrastructures and product design skills in these fields. Unlike the Japanese, they do not focus on making their own manufacturing tools and materials. Instead, they have encouraged tool and material manufacturers to open plants in Korea.

Concerning learning from the United States, Koreans would not repeat the SEMATECH experiment, believing that SEMATECH is an expensive way to orchestrate what the industry should be doing on its own. On the other hand, some in the Korean semiconductor industry hold up the U.S. Semiconductor Research Corporation (SRC) as a model and are trying to set up an SRC equivalent that may facilitate the industry playing a greater role in defining academic curricula in Korea. Korean industry apparently hopes to avoid U.S.-style salary bidding wars for employees by deemphasizing "making money" in exchange for mutual loyalty. Since severe shortages of semiconductor professionals are projected through this decade, whether companies continue to enjoy employee loyalty remains to be seen.

Diversify. Korean semiconductor subsidiaries as well as their parent corporations are very diversified and continue to diversify further. This trend is exactly opposite to the U.S. corporate trend today where spinning off a narrowly focused business unit or merging companies with similar business profiles is very popular for realizing shareholder value (one whose long-term value is a matter of continued national debate). Large Korean companies seem to be protected from the market volatility of narrow business sectors because of their diversity. Diversification, when managed without losing sight of the value provided to customers by each business activity, has so far made Korean companies stronger and more stable.

Globalize and internationalize. These two words are often used together by the spokespersons of Korean companies to describe a part of their growth strategy. This is done to emphasize that going global or international does not simply mean using cheap labor in a foreign country or marketing a product all over the world. Koreans often state that everyone has something to offer, independent of nationality or location. They expect to not only dominate the markets all over the world, but also take their factories and R&D centers all over the world. This idea might appear unrealistic in the competitive markets driven by labor costs. Though difficult to practice, the concept is a valuable strategy for growth.

2.11. INDUSTRIAL COORDINATION — THE KSIA

The concept of a Korean semiconductor industrial cooperative agency might seem a contradiction in terms. Korean electronics firms are fierce competitors, both onshore and offshore. But there are certain common-ground issues that all Korean electronics corporations face. The Korean Semiconductor Industry Association (KSIA) attempts to identify these issues and to act on them on behalf of its members. KSIA is composed of a range of companies that can be broadly grouped in the following categories:

- device technology firms (mostly chip makers and assembly plant operators)
- equipment manufacturers (machinery for assembly, some track development, clean facility tooling)
- materials suppliers (processing chemicals and gases)

The main common concern dealt with by KSIA relates to the fact that Korea is a relatively small country that cannot possibly hope to maintain all the arrays of technology needed to fuel its high-technology ventures. Much technology must be imported through license. KSIA works to ensure that the Korean patent system supports such licensing activity. Korean companies import virtually all of their chip-making equipment from abroad, and they rely on exports to maintain their profit margins. KSIA, therefore, works for favorable trade and tariff policies at home and abroad. The common theme expressed by KSIA officials is that the current electronics economy is global in nature; Korea is not a predator, but a symbiotic component of the world electronics community that produces necessary products (like DRAMS) that other countries could not or would not supply in the required quantities. KSIA works closely with other national and international industry associations (such as the U.S. SIA and AEA, EIAJ of Japan, ECCA of the EU, or SEMI), and it has an associate membership in SEMI and vice versa. It seeks to participate in the major projects of those associations. For example, KSIA and its member firms have joined in the SEMATECH 300 mm wafer effort, in worldwide environmental, health, and safety conferences, and in statistics programs such as WSTC and SICAS. Even though KSIA leadership feels that the Korean industry is too small to significantly influence the direction of such large endeavors, Korean semiconductor makers actively participate and will certainly make use of project results.

2.12. SUMMARY

The future appears to be bright for the Korean semiconductor companies, in spite of the recent downturn in DRAM prices and softening of the semiconductor market that have led the government to nationalize some semiconductor prices and the electronics firms to downsize. Nevertheless, Korean electronics chaebols are poised to take advantage of their own manufacturing skills, facility expansions, and advances in technology. They expect to succeed in taming advanced technologies by outright licensing, participating collaboratively in technology development by providing funding, and by buying companies with advanced technologies. In the past, consumers did not always associate quality with Korean products. However, a shift in perception is occurring because consumers are seeing more Korean products. In particular, Korean computer memories have now gained broad acceptance. More people and companies in the world will contribute to the growth of Korean semiconductor companies either through a direct business relationship or purchase of a product.

The dominance of Korean companies in the international semiconductor component market is clearly acknowledged in the international trade. Japanese government officials announced in January 1996 that Japan will not renew the Semiconductor Component (integrated circuit) Trade Pact with the United States, stating that the U.S.-Japan pact is of little consequence in a global economy and, in fact, is unfair to Japan. They supported their argument by citing that Samsung of Korea is now the leading DRAM manufacturer and that the semiconductor trade is global.

Chapter 3

THE KOREAN DISPLAY INDUSTRY

The Korean display industry has been generally strong in conventional technologies and relatively weak in new technology development; however, it has been rapidly developing new display technologies over the last decade, including the following:

- Black-and-white cathode ray tube (CRT) production was started in 1969. By 1995, Korea had 30% of the world market share of CRTs. The major Korean manufacturing companies are LG Electronics, Samsung Display Devices, and Orion Electronics. Korea is number one in the market based on the number of CRT units sold, but Japan is number one based on earnings.

- In terms of liquid crystal displays (LCDs), Japan is the current leader, but Korea is working to catch up by the year 2000. The major Korean LCD manufacturers are Samsung, Orion, and LG Electronics. The Korean LCD world market share reached 5% in 1995. Current Korean LCDs are mostly TN and STN products.

- There is no plan yet for production of plasma display panels (PDPs) in 1996. Basic research on PDPs, however, has started at Samsung, Hyundai, and Orion, and joint development agreements have been established between Korea (Orion Electronics) and Russia.

- There is no large-scale production of vacuum fluorescent displays (VFDs), although Samsung manufactures them in low volumes. Field emission (FED) and edge-lit (EL) displays are at the basic research level. Samsung, Orion, and Hyundai are the companies conducting exploratory studies.

- A thin-film transfer (TFT) production plan was developed in 1995. The manufacturers are LG Electronics, Samsung, and Hyundai.

3.1. GOVERNMENT PLANNING AND SUPPORT

Strategic planning is not driven by government agencies in the display area; it is only funded and regulated by the government. University professors play key roles in steering new display technology development, in conjunction with individual companies.

The Ministry of Trade, Industry, and Energy (MOTIE), the Ministry of Information and Communication (MOIC), and the Ministry of Science and Technology (MOST) cooperatively support around 40 display development projects, emphasizing manufacturing technology development in information display areas. The government supports six companies on the wide CRT project (16-36 in.).   LCD projects are supported in three subfields: modules, materials and components, and equipment. Fourteen companies are supported through loans and contributions. In the area of equipment research and development, foreign technical groups are participating. Based on interviews with and information supplied by LG executives in June 1995, government support for flat panel displays (FPD) from MOTIE, MOIC, and MOST amounted to $63 million over the last 3-4 years and involved 10 projects.

According to an October 27, 1995, report in the *Chonja Sinmun*, the Korean government has designated the development of advanced flat panel display technologies as part of its G-7 Highly Advanced National (HAN) Projects. The focus of these efforts include LCD, PDP, and FED technologies. MOTIE has also decided to invest $280 million over the five years 1996-2001 in development in these areas. This support will provide both direct financing and tax exemption to participating companies. The projects will include a 27 in. TFT LCD display, a 55 in. PDP, and a 10 in. FED. The government will provide $104.3 million, and private capital will provide $106.3 million. The government will also contribute $35.8 million of a total $72.6 million to be invested in TFT LCD parts and technologies.

In addition, the Ministry of Education will establish short-term FPD university programs to train technical people for industry. Active efforts are also reported underway to pursue technical cooperation agreements, with the assistance of the U.S. Department of Defense. An LCD standards committee also will be established to deal with measurement methods, finished products, parts, and materials.

Further, MOTIE announced on November 30, 1995, that it would designate FPDs as a G-7 HAN project, and $2.7 million would be invested in it. MOTIE's eventual goal is a $7.2 billion FPD production rate by 2001.

Finally, according to MOTIE, Korea has succeeded in developing a 4 in. monochrome FED. A *Chonja Sinmun* article of November 22, 1995, reports that MOTIE and MOIC have been fighting over who funds FED development. It is MOST that makes final decisions on G-7 projects, but individual ministries have responsibilities over specific technical areas. Both MOTIE and MOIC claim responsibility for FED: MOTIE includes

FEDs in the category LCD-PDP projects; MOIC wants to fund FEDs as a single item. The reasoning at MOIC is that FED technology is closely related to semiconductor technology, and because MOIC's Electronics and Telecommunications Research Institute (ETRI) has already done a lot of FED-related research; it is therefore most appropriate for FEDs to be pursued under the direction of MOIC.

As another example of the differences between MOTIE and MOIC, *Chonja Sinmun* also reported that MOTIE plans to develop a full-color 4 in. and 10 in. FED display by 2001. MOIC will be working on critical technologies for a 2 in. monochrome FED display. These contrasting approaches suggest that MOTIE emphasizes production technology and market share, while MOIC emphasizes key fundamental technologies.

3.2. ELECTRONIC DISPLAY INDUSTRIAL RESEARCH ASSOCIATION OF KOREA (EDIRAK)

In the late 1980s, Samsung, LG, Hyundai, and Daewoo approached the Korean government to initiate a consortium to get help in developing flat panel display systems. The Electronic Display Industrial Research Association of Korea (EDIRAK) was approved by the government and established in May 1990. The office is located in the Korean Science and Technology Building in Seoul.

EDIRAK is the Korean equivalent of the U.S. Display Consortium. Regulatory support and some supplementary funding are contributed by the government; the actual sponsors are the member companies. Committee members are responsible for proposal evaluation, adjustment of research directions, and evaluation of progress reports. The function of EDIRAK is to jointly develop and standardize information display technologies through the distribution of government funds to companies for industrial research and development. The research focuses on CRT, LCD, PDP, and FED.

MOTIE provides about 40% of the EDIRAK budget, and companies provide about 60% of the budget, used for office expenses. MOST and MOIC have some role in EDIRAK, mainly in support of the basic research and education aspects of the program. EDIRAK's total funding has been growing; $15 million were distributed in 1995. It is expected that MOTIE will continue to provide major funds in the future.

Funds are distributed in two ways: loans at prime rate, and matching funds (contributions that the industry must match in equal amounts). Funds set aside for loans (which can be 20% of the annual budget) amount to $25 million. Funds set aside for matching funds (which can be 80% of the budget) were $12.5 million in 1995.

About 90% of funded projects are conducted by industrial researchers; 10% of the projects are conducted jointly by industry and university. A university can only submit a joint proposal with industry. Collaboration between industrial researchers and university professors is often based on personal relationships. Usually, professors do not take on administrative aspects of the projects, so the companies manage the administrative responsibilities.

EDIRAK has supported 38 companies. It manages 20-25 projects at any one time, including the loans and contributions. In 1995, it distributed about 30% of its funds to small companies.

EDIRAK proposals are advertised a month before the evaluation takes place. Proposal evaluation takes about a month, and funding begins within two months after the announcement. Proposal evaluation is conducted by a committee of 10 to 20 university professors who generally serve as volunteers. Professors in the evaluation committee do not usually submit proposals to the program in which proposals are being evaluated. Proposals are approximately 15-20 pages. After a company is funded, two progress reports are required each year. Evaluation committees visit the company to monitor the research. Approximately 20% of the grants are said to provide successful research outcomes.

3.3. SIGNIFICANT DEVELOPMENTS

3.3.1 Samsung Electronics Company (SEC)

Samsung (or SEC) has led Korean electronics companies in developing FPD display technologies. Its TFT development history is illustrative of this:

- 1991: organized an R&D team and set-up an R&D line; team size was originally 20 and is now 100
- April 1993: developed a 9.4 in. VGA (512-color FPD)
- September 1993: increased its pilot line to 2400 starts per month, at 300 x 400 mm with 2 panels per glass
- December 1993: started construction on line #1 (contained 20,000 substrates/month)
- February 1994: developed a working sample 12.1 in. XGA with 4,000 colors
- March 1994: developed a low-power (2.6 W) 9.4 in. VGA with 4,000 colors
- June 1994: created a 14.2 in. VGA with 262,144 colors
- September 1994: developed a low-power (2.8W) 10.4 in. VGA with 262,000 colors
- October 1994: completed construction on line #1
- February 1995: started mass production of the 10.4 in. VGA
- June 1995: shipping 9.4 in. and 10.4 in. VGAs; sampling 10.4 in. SVGA (95:Q2) and 12.1 in. SVGA (95:Q4); claims to have demonstrated 10.4 in. VGA FPD using under 2 W of power. Line 1

had 20,000 starts/month with 370 x 470 mm substrates. The equipment cost was $360-400 million (based on approximately 1.2 x the cost of smaller substrate equipment.) This will occupy the first floor of the fabrication facility.

SEC has spent over U.S. $1.32 million on development of a 3.1 in. polysilicon LCD to be used as a key component in a wide range of applications such as movie or slide screens, miniature liquid crystal TVs, video and TV projections, and high-definition TVs. The product is made from polysilicon, allowing for a driver circuit to be integrated into the panel to control the liquid crystal elements of the TFT LCD, thus raising the aperture ratio to 80%. Building the driver IC into the panel also reduces the time needed to fabricate the TFT LCD, reduces the number of production defects, and enhances the reliability of the product. Moreover, the panel size can be reduced to about 2 cm. SEC has completed a TFT LCD mass production line that will enable the company to develop multimedia products as well as to be strongly competitive in international markets.

The new $375 million line includes facilities for processing the substrates used for the TFT LCDs and will ultimately be capable of fabricating the equivalent of up to 80,000 10.4 in. panels (based on a 100% production success rate). The 10.4 in. size is used in notebook PCs, making them the most important market segment. In the first half of 1995, production was expected to be 40,000 panels, with full production (80,000 panels) starting up in the second half of 1995. The company expected to sell $300 million worth of TFT LCDs in 1995, approximately 5% of the world market.

SEC plans to invest an additional $1.25 billion (or more) in R&D and facilities expansion by 2000. By 1997, a second facility will be in production that will have a monthly capacity of 20,000 substrates (or 80,000 panels). By the year 2000, SEC's total capacity will be the equivalent of 480,000 10 in. panels (or 120,000 substrates) a month. SEC also has PDP joint development activities with a Japanese firm.

3.3.2 LG Electronics

LG has joint ventures with Alps Electric in Sendai, Japan, with some technology transfer involved. LG's TFT LCD development history follows:

- 1987: R&D started
- 1990: TFT LCD lab set up
- 1992: prototypes developed
- 1994: pilot manufacturing started
- June 1995: mass production fabrication facility completed

3.3.3 Orion Electric

Orion Electric, owned by the Daewoo group, makes various display products. Approximately 85% of its $1 billion of 1995 sales was in color picture tubes and color display tubes. Other key developments include the following:

- The company's 70:30 $170 million joint venture with the Hanoi Electric Corporation in Vietnam, Orion-Hanel Picture Tube Co., Ltd., turned out its first color picture tube in July 1995. Orion-Hanel has an annual production capacity of one million color picture tubes (CPT) for color televisions and 600,000 black-and-white picture tubes for televisions monitors. Orion-Hanel made its maiden export, 25,000 CPTs, to Indonesia in October 1995 and has sent additional shipments to Thailand and Brazil. The plant is also meeting Vietnamese domestic market requirements for color television production.

- Orion Electric holds a 15% equity interest in a new $110 million project, P.T. Tosummit Electronic Devices Indonesia, with Toshiba and Sumitomo of Japan (combined 50% interest) and P.T. Tabung Gambar Indonesia (35% interest) as partners. The companies held a ground-breaking ceremony for a new CPT plant in West Java that will have an annual production capacity of 2.3 million CPTs by 1999.

- Orion Electric and Daewoo Corporation will hold a combined 50% equity interest in a joint venture CPT plant with the Anglo American Corporation (AMIC) of South Africa. The companies plan to construct a new production facility in Johannesburg, which will meet both domestic and export demands for CPTs.

- Orion Electric is a 50:50 joint-venture partner with Daewoo Electronics in Daewoo-Orion Société Anonyme, in France, which produces 1.2 million CPTs annually. It also plans a CPT production plant in San Luis Rio Colorado, Mexico, and it has a sales subsidiary project in San Mateo, California.

- In 1995, the company made a total investment of $163 million into research and development. As a result, it introduced a number of new products, including a 28 in. wide vision cathode ray tube (CRT); a 17 in. color monitor; a 10.4 in. color monitor STN-LCD for notebook PCs; and a 21 in. color plasma display panel (PDP) for televisions.

- In 1995, Orion Electric signed a cooperative agreement with Toshiba of Japan in the field of liquid crystal displays. Through the agreement, Orion Electric will produce super twisted nematic (STN) LCDs, and Toshiba, which has substituted its STN LCD production with thin film transistor (TFT) LCD production, will market Orion Electric's STN LCDs globally through its network. Orion Electric began mass production of 10.4 in. STN LCDs in October.

- Orion Electric, as part of Daewoo Group's growing emphasis on industrial-academic cooperation, has established a new Orion-Ajoo University FED Research Center during the year to concentrate on development of FEDs. Orion Electric has set the ambitious goal of becoming one of the world's top three specialty manufacturers of displays by the year 2000.

- Orion Electric entered into a 50:50 joint-venture partner with a Russian firm in the Orion Plasma Research and Production Company of Ryazan, Russia, for the development of PDPs, 100 in. and 200 in. multi-screens, and other displays.

3.3.4 Hyundai

Hyundai's TFT LCD plan as of June 1995 was to spend $428 million from January 1995 to June 1996, of which $151 million was to fund construction and $277 million was to fund production equipment. The production plans were to have Line 1 at 20,000 starts per month at 370 x 470 mm. Phase 1 was planned for March 1996, for production of 6,500/month, and Phase 2 was planned for October 1996, for production of 13,500/month. Since 1992, Hyundai has had a joint venture with Imagequest in the United States: it acquired technical material but also invested $20 million in the venture. Imagequest has about $3.2 million in annual sales, primarily in avionics displays, with its high-end products being used in the Boeing 777 and F-16.

3.4. SIGNIFICANT RESEARCH EFFORTS

The electronics industry consortium EDIRAK, under the auspices of MOTIE, created the Next-Generation Litho Display Technology Development program in order to develop 25-29 in. high-quality, low-electric-power TFT LCDs (XGA-Standard 12 in., Level 1.3 W), and wall style 55 in. full-color PDPs for HDTV. The total budget of the project is over $235 million, covering a research and development period from 1995 to 2001. The large companies combined have invested slightly more of the total research budget than the small- to medium-sized companies.

The program has three major projects (Tables 3.1-3.18): development of a 25-29 in. high-quality TFT LCD; development of a low-electric-power TFT LCD; and development of a full-color 55 in. PDP flat-style HDTV.

3.4.1 Project for Development of a 25-29 in. High-Quality TFT LCD
This project includes the following elements (see also Tables 3.1-3.3):

- Forecasting the use of litho-display technology, including development tendencies, market analysis and tendencies, and environmental analysis
- Development of related technologies for a 25-29 in. a-Si TFT LCD, including wide visual angles, large scale, lower-cost TFT

manufacturing, high opening rate (mobility/BM on array), improved screen quality via improved surface reflection, and design/simulation

- Development of 25-29 in. a-Si TFT LCD parts and equipment, including development of color filter, backlight unit, in-line coaster, ion shower equipment, and etching system
- Basic technology research for 25-29 in. a-Si TFT LCD, including wide visual angle enfilade film, and high-speed mobility, TFT, and Matrix

Table 3.1
Project for Development of High-Grade 25-29 in. TFT LCD (a)

Sub Project	Forecasting technology in Litho-display
Unit Project	Forecasting technology in Litho-display
Goals	Technology analysis of Litho-display
	Forecasting market size
	Application analysis
	Direction of technology bottle-necks and breakthroughs
Key Research Items	Present global R&D conditions for important LCD, PDP and FED advances in the Litho-display area
	Detailed assemblies and technological systems
	Newly released application products with forecasts of domestic and foreign market trends
	New technology trends
	Application areas of each kind and size for Litho-display
	Obstacles and pace for development of technology and standardizing of domestic technological development

Table 3.2
Project for Development of 25-29 in. TFT LCD (b)

Sub Project	Basic research of large-size a-Si TFT LCD
Unit Project	Basic research of wide visual angle/surface enfilade film
Goals	Technology of wide visual angle by low cost processing (140/140)
	Technology of wide visual angle without damage in TFT
	Technology of wide visual angle without color shift and V shift
Key Research Items	Basic research of wide visual angle/surface enfilade film
	Development of wide visual angle technology by new method (obtaining a patent)
	Process control technology without damage in TFT
	Process control technology applied to large sizes

Table 3.3.
Project for Development of 25-29 in. TFT LCD (c)

Sub Project	Basic research for large size a-Si TFT LCD
Unit Project	Basic research for high mobility TFT and matrix
Goals	Technology for high mobility a-Si TFT (3 cm^2/Vs)
	Technology for advanced matrix TFT LCD (simple process, wide visual angle, large size)
Key Research Items	Basic technology for TFT, TFT matrix, TFT array suitable to produce large size TFT LCD
	Technology for high mobility a-Si TFT
	Research for weakened capacitance minimization, self-alignment, TFT particularization

3.4.2 Project for the Development of a Low-Electric-Power TFT LCD
This project includes the following elements (see also Tables 3.4-3.5):

• Development of a low-electric-power a-Si TFT LCD, including development of BM on array technology, of high-opening rate technology, and of simple structure matrix technologies

• Development of a low-electric-power poly-Si TFT LCD, including development of laser Poly-Si TFT array technology, and of processes for doping insulation plates

• Development of a color filter with high opening rate, a high-efficiency invertor, high-efficiency backlight unit, and high-efficiency flat light

• Basic research for low-voltage Ld and for low-temperature poly-crystallization silicon TFT/matrix manufacturing

Table 3.4
Project for Development of Low-Electric-Power TFT LCD (a)

Sub Project	Basic research for low electric power
Unit Project	Basic research of producing low-temperature polycrystalline silicon TFT/Matrix
Goals	Technology (>400 cm^2/Vx) for low-temperature (below 400°C) polycrystalline silicon
	Technology of new frame TFT, TFT matrix, and TFT array (applicable to low-electric-power LCD)
Key Research Items	Technology for high-mobility, low-temperature poly-crystalline silicon TFT
	Research of poly-Si TFT applicable to low-electric-power TFT LCD
	Technology for new driving circuit for TFT LCD

Table 3.5
Project for Development of Low-Electric-Power TFT LCD (b)

Sub Project	Basic research for low electric power
Unit Project	Basic research for driving liquid crystal (LC) with low electric pressure
Goals	Basic research for low-electric-power TFT LCD (liquid crystal with low electric power, LC mode without polishing plate)
	Useable LC technology to produce XGA (12") TFT LCD
Key Research Items	Basic research related to LC low-electric-power TFT LCD

3.4.3 Project for Development of a 55 in. Full-Color PDP Flat-Style HDTV

This project consists of developing element and production technology to the following specifications:

Size:	55 in.
Brightness	250 cd/m^2
Efficiency:	1.0 lm/W
Durability:	Over 20,000 Hr
Electric Usage:	180 W
Contrast Ratio:	50:1
Gray Scale:	256 Full Color
Visual Angle:	160
Pixel Pitch:	Less than 0.5 mm

For the element technology, the research consists of the following:

- Investigating the high-speed driving circuit for HDTV, with low-voltage and driving IV, reaching a high gradation, and achieving a high brightness
- Focusing on obtaining a high efficiency, high degree of color purity development, and fluorescent plate formation technology
- Researching feed-through formation technology for high fixed and large size
- Analyzing high-brightness, high-efficiency, and long-life discharge cell design technology
- Examining discharge current control resistible plate material and formation technology
- Exploring electrode material and electrode formation outlining the low-resistible electrode material and electrode formation, transparent electrode development (low-resistible, high-transmissivity) formation technology of electrodes with excellent inner sputtering characteristics

- Studying the shield plate and dielectric substance, focusing on technologies to form large-size shield plates and dielectric plates
- Optimizing the electric discharge gas
- Processing technology of large plate glass
- Developing technologies to form large thin-film and electrode technology to form large, thick film
- Producing an electrode arrangement technology with feed-through minimizing, and producing highly refined resistance plate and fluorescent plate technology
- Creating a technology for in-line process of exhaust and sealing equipment
- Developing technology for sputtering and sand blast equipment

3.5. SUMMARY

In the late 1980s, the top four Korean companies decided that they needed to be able to design and manufacture displays in order to be competitive in the future world electronics marketplace. They have clearly succeeded in getting their foot in the door — an achievement that has eluded U.S. consortiums, despite heavy funding. If the Korean companies follow on the path of their success in DRAMs, then their future in the display arena looks exceptionally good.

Chapter 4

PRODUCTS AND SYSTEMS

While most highly noted for its successes in the semiconductor memory market, the Korean electronics industry has also been a significant player in the world markets for personal computers, telecommunications equipment, and automotive and consumer electronics products. These activities indicate a broad-based standard of activity and a balanced posture in the global electronics community.

This chapter overviews the wide variety of products and systems being developed by Korean companies. The focus is on the information services, from computers to satellite communication technologies.

4.1 COMPUTERS

Korean companies have begun developing various multimedia computer technologies to enable computers to manipulate voice, image, video, and graphics. In July 1990, a national R&D project was started in order to develop a multimedia computer. In the first phase (July 1990 - July 1994), a multimedia PC (ComBi Station) and personal multimedia workstation (ComBi Station I) were developed. In the second phase (July 1994 - June 1998), an intelligent multimedia workstation has been under development.

Utilizing its own multimedia processing software with the assistance of proprietary hardware, the ComBi Station executes multimedia applications in real time. It also provides users with an agent-based user interface, which is a new concept of an intelligent interface that could be adopted in next-generation computers. The ComBi Station is operated under Microsoft's Windows NT. It provides client-server features and therefore allows connecting to NAIS host computers. ComBi Station I is a high-performance personal workstation aimed at high-end application markets. An electronic multimedia book (Oksuh) and a multimedia authoring tool (Okdang) have also been developed.

Following the successful development of TICOM II and III, a joint R&D project for a state-of-the-art highly parallel computer, called TICOM IV, is being carried out. TICOM IV is a four-year project that was initiated in February 1994 and will continue until January 1998. The overall cost estimated for the project is about $70 million, half of which is being invested by the Ministry of Information and Communications (MOIC) and the Ministry of Science and Technology (MOST), and the other half by participating companies, including Samsung, LG, Daewoo, and Hyundai, and by research institutes. The overall manpower invested is expected to be 570 man-years.

TICOM IV uses Intel P6 chips as its main processors, and it can be configured with a maximum of 256 processors, scaleable processing architecture, and an open-system architecture that satisfies international standards. TICOM IV has a 20 GIPS data processing capability and 10,000 tpmC transaction processing capability. This computer is targeted for application in a world-leading business parallel system; it will be also used in various information and communications services of the future information society. TICOM IV will be introduced in the marketplace in 1998.

R&D for a gigabit information processing and networking technology (GIANT) began in 1993 to provide, among other things, distributed multimedia processing, communication device agents, tools for multimedia database build-up and information retrieval, and a multimedia communication protocol. The object of the GIANT project is to establish an environment enabling "anywhere, anytime, and anyhow" types of information networking services.

4.2 MULTIMEDIA COMMUNICATIONS

The Korean focus on multimedia communications includes HDTV, video conferencing, and spoken language processing. Based on the MPEG-2 CODEC chip set that can process video and audio signal compression, Korean companies and broadcasting firms have developed a domestic standard for an HDTV transmission system. They have also developed a video terminal adapter for broadband integrated services and data networks (B-ISDN), a desktop video conferencing terminal for ISDN (in 1994), and are now developing optical cable TV with video-on-demand service.

In 1995, the Electronics and Telecommunications Research Institute (ETRI) developed a Korean-Japanese interpretation system for hotel reservation tasks. ETRI has also designed a spontaneous speech translation system called "Speechmate" to handle male and female voice with good prosody. These systems are already being transferred to Korean companies.

4.3 COMMUNICATIONS PROCESSING TECHNOLOGY

The Information Communications Processing System (ICPS) was developed to act as a gateway for information services. It performs as an interconnect between user terminals in public switching telephone networks (PSTN) and an information service center in a public switching data network (PSDN). It allows users to have open information retrieval capabilities, value-added network (VAN) services, and information guidance services without repeated registration or subscription processes. It underwent a commercialization test in 1993 and was field-tested in August 1994. After successfully completing test operations in six large cities, ICPS went into operation in March 1995.

ICPS consists of information service access equipment called Information Service Access Points (ISAPs) and maintenance/operation systems called Mediation-Operation Administration and Maintenance (M-OAMs) systems. Each M-OAM is constructed by a multiprocessor computer called TICOM (see above section on computers).

As to the adaptation of B-ISDN to ISDN, LAN, video and frame relay, four different pieces of Broadband-Terminal Adapter (B-TA) equipment were developed:

- ISDN B-TA for ISDN voice and image telephones
- LAN B-TA for 10 Mbps-range Ethernet
- Video B-TA for DS3-level coded (45Mbps) NTSC images
- Frame relay B-TA for DS1E-level (2 Mbps) frame relay networks.

Development of these types of B-TA equipment started in 1992. The first product was produced in the middle of 1994, and additional products were completed in 1995. The B-TA equipment can easily allow adaptation to existing network services, which in turn is expected to lead the evolution of asynchronous transfer mode (ATM) networks and high-speed communications systems. As a result, B-TA equipment can vitalize various high-level communications services expected to emerge in the near future.

These developments are used in the Korean multimedia market; however, it is too early to tell what the global impact will be.

4.4 TRANSMISSION TECHNOLOGY

Transmission technology is essential to future B-ISDN. ITU-T, in particular, has announced Synchronous Digital Hierarchy (SDH) that adopts 155 Mbps as a basic signal to form Synchronous Transport Module-level N (STM-N; N = 4.16.64) signals.

ETRI has been developing SDH-based systems that interface 155 Mbps, 622 Mbps, and 2.5 Gbps optical signals. The 1994 results of R&D for transmission technology include synchronous multiplexer developments based on STM-1 optical signal interfaces. Two simple terminal multiplexers were developed in 1992 and 1993, respectively: the Synchronous Multiplexer & Optical Terminal, level -1 (SMOT-1), and the Add/Drop Multiplexer-155 (ADM-155). Commercialization feasibility was assessed by field tests. In particular, the network configurations for field testing formed PTP (point-to-point) connections between two SMOT-1 pieces of equipment at different telephone offices, and ADM rings using ADM-155 equipment at four sites based on the unidirectional path switched ring protection scheme. Currently, commercial versions of both types of equipment are in mass production by four Korean telecommunications companies.

A 2.5 Gbps optical transmission system called HLS-2500 was developed from 1988-1992. Its basic technology was transferred to domestic companies in 1993, and in 1994, those companies implemented the systems for verification testing. Each company has implemented its own commercial prototype for development as an optical link between Seoul and Taejon.

The HLS-2500 system can be used as a terminal multiplexer, linear ADM, and ring ADM, according to the location of the system in the network. The system can interface both PDH (DS3) and SDH (STM-1, STM-4) signals as tributary signals. The system's size was minimized by applying ASIC technology to logic designs. System reliability and availability have been improved through appropriate adaptation of system maintenance and TMN functions.

The development of BDCS began in 1992. There are two versions of BDCS. Based on the SDH multiplexing procedure, BDCS I, which provides cross-connection functions on 50 Mbps and 150 Mbps path signals, will be developed before the second version, BDCS II, which will provide additional switching function on 2 Mbps signals. BDCS can interface up to ten 2.5 Gbps optical signals for the cross-connection of 2 Mbps, 50 Mbps, and 150 Mbps multiplexed in those signals. It may also perform add/drop functions on DS1N, DS1E, and DS3 PDH signals with a maximum capacity of five 2.5 Gbps signals. It will perform three protection functions: span switching protection, Self Healing Ring (SHR) protection, and Self Healing Mesh (SHM) protection, based on the network topologies.

The BDCS I laboratory prototype was developed in 1994. It consists of a digital cross-connect subsystem (DIXS) that cross-connects 50 Mbps signals with a 576 x 576 switching matrix, an add/drop signal interfacing subsystem (ASIS) that provides add/drop functions on multiplexed PDH signals (DS1N, DS1E, DS3 signals), a BDCS operation and surveillance subsystem (BOSS) that includes more than ten 32-bit processors. To minimize system size, three ASICs of 20-30,000 gate sizes are being designed. To make the BDCS I system serve as a hub node in the network, an SDH network management function is being developed. Since August 1994, three domestic companies have joined the system development effort.

An optical power amplifier/preamplifier prototype was developed for the 10 Gbit-rate long-haul optical test. In 1994, a study of 10 Gbit-rate high-speed multiplexing and maintenance functions was conducted. To realize the multiplexing function, ASIC design is in progress. The first prototype system was planned for implementation in 1996. It is not known at this time who will be manufacturing the ASICs.

POTS, an N-ISDN interface system, a broadband switching system, and subscriber equipment have all been developed as main technologies for the optical CATV system. Development began in 1994 by several Korean companies, with subsequent studies to investigate the feasibility of using TDM/Polling technology to enhance the switching control system.

4.5 BROADBAND INTEGRATED SERVICES DIGITAL NETWORKS

Most major systems to be used in constructing the information superhighway in Korea will be under development until 1998. The system technologies being developed within the government-sponsored Highly Advanced National (HAN) B-ISDN project are expected to be equivalent to those of advanced countries by the year 2000. The HAN B-ISDN project was started in order to develop network integration technology, ATM switching, 10 Gbps and 100 Gbps transmission systems, a broadband network termination system, a broadband terminal adapter, and ATM terminals to support the information superhighway.

Many broadband network systems are now commercially available in both private and public networks. However, it is expected that B-ISDN will be an infrastructure for construction of various information network services such as video-on-demand, multimedia distribution, and retrieval services for the information superhighway. Cost-effective fiber deployment technology (Fiber to the Curb, FTC) will be used to provide users with a bandwidth of 155 Mbps.

4.6 INTELLIGENT NETWORK TECHNOLOGY

Development of an "Intelligent Network" (IN), regarded as one of the most innovative and evolutionary architectural concepts, is accelerating the introduction of new, advanced telecommunications services in Korea. In its "IN Service Systems Development Project" (1988-1994), ETRI has developed Common Channel Signaling (CCS) equipment, intelligent network service controlling and management equipment, and various service logics in order to realize the Intelligent Network. SMX-1 is a major CCS equipment item used for signaling transfer points. SMX technology has been already been transferred to equipment manufacturers for commercial production. SMX/CMP, which is a smaller version of SMX-1, has also been developed to be used as a CCS No. 7 front-end processing unit for various signaling points such as SCP. Development of SIGNOS, a signaling network operations and management system, was completed in 1993.

The goal is to transform the current channel-associated signaling network into a more intelligent common-channel signaling network. NICS (Network Information Control System) is a service control and management platform, realized as a realtime fault-tolerant multiprocessor system. Currently, two services — Freephone (a Premium-Rate service), and credit calling — have been implemented on the NICS platform. An advanced Intelligent Network is also being studied.

4.7 ATM SWITCHING SYSTEMS

Through their implementation of the nationwide B-ISDN project, ETRI, MIC, Korea Telecom, industry, and academia have scheduled the completion in 1998 of a large-scale commercial ATM switching system. In 1994, a 64 x 64 experimental ATM switching system was developed and tested. The main processor board, based on a high-performance CPU, was designed in 1995. In the implementation of the ATM switch, a 16 x 16 switching element with shared common memory was designed and developed, as well as a 188 Mbps switch network link, a control unit for testing and diagnosing switching elements, and a network synchronization unit that generates and distributes various clocks needed in the ATM switching operation. In 1996, the focus was on improving the present ATM switching model to include more intelligent functions.

In the switching elements area, ETRI researchers developed ASICs for bit synchronization of the transmitter/receiver of the ATM switching system for 155 Mbps subscriber interfaces; they also designed 622.08 Mbps bit synchronization ICs. In 1996, they planned to devise a 622 Mbps transmitter and receiver. Also planned are the development of a 622 MHz call synchronization IC and ASICs for ATM application. A logic design for a photonic switch and a SDH IC and a simulation for the 155 M SOH and the POH IC have been developed. Researchers have also worked out the logic design and a simulation for a 622M SDH IC. They demonstrated a newly structured 2 x 2 laser amplifier gate switch matrix with a very low fiber-to-fiber loss.

4.8 WIRELESS ISDN SERVICE

The Wireless ISDN possesses intelligent network switching and radio interface capabilities and provides mobile switching calls to mobile users at the same service levels as those of PSTN/ISDN subscribers. Research in this area has focused on architecture, performance evaluation, services and standardization, and test environments for switching systems. The principal results include the following:

- switching system requirements were delineated, as well as architecture based on the TDX-10 ISDN/SSP
- performance objectives for switching call processing with mobility modeling were outlined and subscribers' traffic analysis carried out
- a call processing sequence chart was developed
- user functions for mobile call processing services were analyzed, as were the Personal Communication Service standards and services of advanced countries.

4.9 TDX-10 ISDN

The goal of ETRI's TDX-10 ISDN project is to develop a switching system that can accommodate various non-voice services and advanced switching technologies for the next-generation broadband switches. The project began in 1991; commercial tests were being conducted by 1994.

In parallel with development of the ISDN switching system, ETRI has also developed ISDN terminals and ISDN chips. It has completed national standardization for ISDN terminals, including an ISDN telephone set, a PC S-interface card, and TA (terminal adapter). Also, supplementary development of complex multifunctional Desktop Videoconference Terminals (DVT) was carried out, and core technologies transferred to collaborating companies. ISDN chips, including two kinds of subscriber chip (U-interface digital IC and U-interface analog IC) are still being developed.

ETRI is actively collaborating with UTA (University of Texas at Arlington) on object-oriented testing technology in the development of a new switching system, drawing on its broad experience accumulated in the TDX-10 ISDN development project. The vision for this new switching system is that it will be sufficiently attractive in terms of pricing, performance, and switching functions to be competitive with foreign switching systems.

4.10 DIGITAL MOBILE COMMUNICATION SYSTEMS

Mobile telecommunications technology, which has played a leading role in moving society into the information age, has been able to fill skyrocketing demand for mobile telecommunications products and services through the introduction of newer core technologies such as frequency reuse and microcell technologies. The inadequate capacity of the analog cellular mobile system requires deployment of a digital cellular mobile system.

There are two principal methods considered feasible for second-generation digital mobile telecommunications using current technology: Time Division Multiple Access (TDMA) and Code Division Multiple Access (CDMA) methods. TDMA was developed and commercialized in Europe's GSM and Japan's PHP systems. CDMA was proposed by QUALCOMM, Inc., of the United States, and is being developed with several Korean manufacturers, telecommunications societies, and research institutes with the aim of producing a fully integrated system in a short period of time. Because of its high capacity (10-20 times more than AMPS) and its benefit of privacy, CDMA has been favored for use in the Personal Communication Network (PCN) system planned to be implemented soon.

Launched in 1989, this project successfully developed and tested a commercial CDMA system in 1994. In 1996, the major development activities were to be the operational testing of the integrated system and the development of ASICs. Commercial services were to be deployed in 1996. R&D is now focusing on new technologies such as Personal Cellular System (PCS), mobile-satellite telecommunications, and Future Public Land Mobile Telecommunications Systems (FPLMT). Korean manufacturers are now developing a CDMA system for second-generation digital mobile telecommunications.

4.11 PERSONAL COMMUNICATIONS SERVICES (PCS)

The objectives of PCS are ease of use, choice of price, and means of communicating with anyone, anytime, anywhere — whether at work, at home, or on the move. Several countries have implemented PCS systems of their own to preempt the PCS markets. In Europe, GSM- and DCS-1800 based systems have already been developed and are in service. In Japan, the PHS system was developed. In the United States, the Joint Technical Committee (JTC) a joint meeting between T1P1 and TR46, has been taking the initiative in standardizing radio-interface PCS.

Korean telecommunications carriers have been investigating various aspects of PCS. Their main R&D activities include planning PCS development programs, organizing technical groups, and studying radio interface technologies. ETRI began developing a PCS system in 1995 in order to acquire the core radio interface and radio-related MMIC and ASIC technologies.

4.12 SATELLITE COMMUNICATIONS TECHNOLOGY

During the last three decades, Korean satellite communications technology has developed rapidly. Launched by a McDonnell-Douglas Delta II rocket in 1995, the first-generation Koreasat opened a new era of satellite communications in Korea. Martin Marietta of the United States was the primary contractor and Matra Marconi of England was the

secondary contractor.   Thirty Korean engineers were sent to those
companies for the on-the-job training.   The DAMA/SCPC system and the
VSAT system were successfully developed in a joint effort that included five
Korean companies and two foreign companies.   These systems were tested
and operated through a leased satellite, Intelsat VII.

The current development goals of satellite communications are
technical support for construction of the Koreasat network, domestic
development of earth station facilities such as DAMA/SCPC and VSAT,
and development of a digital satellite broadcasting transmission system
expected to be operational in 1996.

4.13 SUMMARY

The development of electronics technologies in Korea is broad-based in
the sense that it is not linked to a single product or product set.   Korean
companies appear to be strategically oriented towards developing a broad-
based, balanced posture in the global electronics marketplace.   For example,
Korean companies have expanded into multimedia, telecommunications,
internet, and a wide range of associated information technologies.   These
activities have already been integrated into the Korean marketplace, and the
impact on the global level should be evident in the late 1990s.

Chapter 5

KEY KOREAN ELECTRONICS COMPANIES

This chapter profiles some of the key "giants" of Korean electronics, including basic information on their product lines, personnel complements, and the authors' understanding of individual company strategies and philosophies.

5.1 SAMSUNG ELECTRONICS CORPORATION

Samsung Corporation was founded in 1938 in Taegu, Korea, as a trading firm to supply rice and agricultural commodities to neighboring countries. In 1948, the company moved to Seoul and broadened the focus of its operations by adding partners throughout Southeast Asia and in the United States. From 1950 onward, company growth has been exponential. Revenues since 1987 have tripled.   The 1994 annual revenue was $64 billion.

Samsung has branched into wood and textile products (1954), electronics (1969), construction and petrochemicals (1974), aerospace (1979), commercial vehicles (1994), and passenger cars (1995). Samsung Electronics Corporation has further subdivided: the Semiconductor Division was founded in 1974 and the VLSI Division was founded in 1982.

5.1.1 Current Organization

The company is now divided into five main subgroups: electronics, chemicals, machinery, finance, and insurance. Significant measures of success are evident in a number of these business units. In the area of electronics, despite a relatively late start in the field, the Electronics Division was the world's leading supplier of black and white televisions in the early 1970s. What is known as Samsung's Electronics Subgroup is now the world's largest supplier of dynamic random access memories (DRAMs) and the world's seventh largest semiconductor device manufacturer. It employs over 77,000 people. The subgroup has seven major affiliates:

Samsung Electronics Co., Ltd., Samsung Display Co., Ltd., Samsung Electro-Mechanics Co., Ltd., Samsung Corning Co., Ltd., Samsung Data Systems Co., Ltd., Hewlett-Packard Korea Co., Ltd., and Samsung-GE Medical Systems Co., Ltd. The large number of affiliate units (including foreign alliances) reflects the trend to decentralized, employee-oriented management practices. Semiconductor device production takes place within the Samsung Electronics Corporation (SEC), the focus of this summary, although background data is supplied for other subgroup activities as well.

5.1.2 Products (by Affiliate)

- *Samsung Electronics Corporation.* SEC is involved in the full range of electronics-related systems and products, including semiconductors, information systems, computers and consumer electronics. SEC was the first to develop a fully working die of the 256 Mbit DRAM and is Korea's leading electronics maker in global markets. Promoting a high-tech image, SEC achieved $21 billion in total sales in 1995, up 40% from the 1994 figure of $14.6 billion.

- *Samsung Display Devices.* Samsung Display Devices (SDD) is the world's largest producer of Braun tubes. SDD is also diversifying into high-value-added picture tubes such as ultraflat and wide-screen models, as well as next-generation display products to include super-twisted nematic LCDs, vacuum fluorescent displays, and light-emitting diodes.

- *Samsung Corning.* Samsung Corning is a joint venture with Corning, Inc., of the United States. Main products include glass for TV picture tubes, glass for liquid crystal displays (LCDs), soft ferrite, and ceramic filters. Recently the company has been expanding its operations in Malaysia and Germany, and by 1997 Samsung Corning will be the world's third-largest TV glass producer, with a total annual capacity of glass for 50 million picture tubes — 25% of the world market.

- *Samsung Electro-Mechanics.* Samsung Electro-Mechanics specializes in parts and components for electronic products such as deflection yokes and flyback transformers. The company has developed the world's smallest tuners (15 cc) as well as chip-type electrolytic condensers. With the establishment of Samsung Motors, Inc., Samsung Electro-Mechanics is also expanding into the automotive parts business. Five percent of total sales is now being invested in R&D as the company strives to become one of the world's top three parts-makers by the year 2000.

- *Samsung Data Systems.* Samsung Data Systems (SDS) is Korea's largest information services company and is involved in systems integration, systems management, telecommunications services, and training.   SDS is responsible for the Samsung Group's overall computerization, including standardizing the various information systems for each business area and job function and constructing a group-wide information network. The company opened the Samsung Information Academy in November 1994 to provide employees and management group-wide specialized courses in computing. The academy also offers training to the general public.

• *Foreign Ventures.* Hewlett-Packard Korea Co., Ltd., and Samsung-GE Medical Systems Co., Ltd., are largely outlets and assembly points for Hewlett-Packard electronics instruments and for GE medical equipment (such as X-ray diagnostic tools).

5.1.3 Research & Development

In early December of 1995, SEC announced prototyping results of the worlds first 1 Gbit DRAM. This represents the major development effort of the company. Two points should be emphasized regarding this significant achievement: First and foremost is the fact that DRAM fabrication is considered to be the "cutting-edge" technology for the semiconductor industry. This implies that SEC is now practicing the highest level of technology in what is probably the most technologically demanding area of the industry today. Also, Korea has significantly advanced the state of the art in three ways:

1.   Minimum feature sizes in this device were reported to be 0.18 μm, and the base design-rule was consistent with 0.2 mm technology. Current process capability worldwide supports 0.25 μm design rules in prototyping.   The lithography employed by SEC was the most aggressive application of optical systems to date: deep-UV 248 nm off-axis quadruple illuminators were integrated with half-tone phase-shift masks and chemical-mechanical polishing (CMP).

2.   The individual memory cell was reengineered to create a more planar structure to ease the burden on the lithographic process. The resulting cell is referred to as an Isolation Merged Bitline Cell (IMBC).

3.   Overall system architecture was optimized for low-power application and new redundancy schemes were employed to fix failing bits.

It is worth noting that all of these accomplishments were based on an imported tool base. Thus, other corporations and other nations have "equal access" to the processing capabilities employed. Korea has no indigenous semiconductor equipment industry and thus must rely on foreign sources for tooling. The key device for achieving the 1 Gbit DRAM demonstration was a U.S. stepper — an SVG-L microscan system.

SEC representatives were reluctant to talk on technological issues. The company's position is that it has adequate cooperative ties through established agencies (such as the U.S. Semiconductor Industries Association, SIA). Advanced VLSI technology is highly competitive, and long-range plans and approaches in this area are "company proprietary." There was a strong feeling expressed that success in the DRAM market has been the result of smart business decisions on the part of the company's founders and directors, and that external assistance from the government or through interindustry research efforts is unnecessary and could compromise the company's position. VLSI contains very little in the way of "precompetitive" development requirements: government assistance may be important for small, emerging companies, but not for a giant like Samsung.

Overall, SEC invests about 5% of its total sales on R&D. It now employs over 11,000 workers in the research area. Samsung plans to invest 12% of its total sales in R&D by the year 2000, and to employ over 40,000 researchers. Samsung's Advanced Institute of Technology (SAIT, established in 1987) is the company's main research arm. It addresses eight focus areas: digital signal processing, laser applications, displays, mechanical/electronic interfaces, computer software, data transmission, new materials, and environmental protection/alternate energy sources.

5.1.4 Government and University Interactions

Just as the other major Korean electronics corporations surveyed, Samsung managers do not directly attribute any of the company's success to government support; rather, they see their success as the result of the foresight and leadership of their founders. Of course, universities are important as sources of skilled workers. The company does not look to the university as a source of new product ideas. This follows along with a more-or-less commonly held Korean belief that universities are essential, but only in so far as they provide talented personnel. Even the best graduates require a 6-12 month course of study at in-house "technology universities" to acclimate them to the new work environment.

5.1.5 Alliances

Currently, Samsung maintains a number of joint ventures and technical cooperation agreements with foreign agencies. SEC's major alliances and joint ventures are summarized in Tables 5.1 and 5.2. Primary foreign alliances are with Japan. It was through these alliances that DRAM production was initiated in Korea. Only recently (1993) has Samsung begun selective collaboration with a U.S. vendor (Micron Technology) in DRAM production. Samsung maintains no ventures or collaborative efforts with other Korean electronics firms.

Table 5.1
SEC's Major International Technical Alliances

Partner	Year	Scope and Content
Dancall (Denmark)	1995	Joint development of DCS 1800/GSM mobile phone system
Toshiba (Japan)	1995	Joint development of ICs for use in consumer electronics products
Fujitsu (Japan)	1995	Technology sharing for sophisticated next-generation TFT LCDs
Hewlett-Packard (U.S.)	1990	Joint development and sales of RISC workstations: HP to provide technical assistance; SEC to develop/manufacture
Toshiba (Japan)	1992	Technical cooperation/flash memory devices
IBM (U.S.)	1993	Joint development and sales of desktop PCs: IBM to market and sell; SEC to manufacture
OKI (Japan)	1992	Technology transfer and technical assistance for synchronous DRAMs
Mitsubishi (Japan)	1993	Cooperation in cached DRAMs
Micron Technology (U.S.)	1993	Cooperation in next-generation memory devices
AT&T (U.S.)	1993	Joint development of notebook and pen-based PCs
General Instruments	1994	Joint development and sales of HDTV: GI to provide technical assistance; SEC to develop and manufacture

Source: Samsung World Wide Web site

Table 5.2
Joint Ventures, Samsung Business Unit and Foreign Corporations

Partnership	Year	Scope and Content
SEC/DNS (Japan)	1992	Joint development and production of semiconductor manufacturing equipment
SEC/NEC (Japan)	1995	Produce semiconductors for EU market
SEC/Texas Instruments (U.S.)	1994	Semiconductor production at TI's existing plant in Porto, Portugal
Samsung Corning/Corning (U.S.)	1973	TV picture tube glass production (50/50 joint venture)
Samsung-GE Medical Systems/General Electric (U.S.)	1984	Medical diagnostic equipment
Hewlett-Packard Korea/Hewlett-Packard (U.S.)	1985	Production and marketing of computers, instruments, medical diagnostic machines

Source: Samsung World Wide Web site

5.1.6 Other Overseas Activity

Of further interest is the degree of foreign development planned. SEC just broke ground for a new $100 million assembly plant in Suzhou, China (PRC). The spin-off division, called Samsung Electronics (Suzhou) Semiconductor Co., Ltd. (SESS), is targeted to be Samsung's first site for production of nonmemory devices. As discussed above, the road to DRAM production in Korea was through alliances with the Japanese. It might be envisioned that Korea would pursue ASIC or random logic designs through affiliations with the United States. Collaborations with Micron Technology are underway, and IBM already maintains technical agreement portfolios with Samsung in the area of personal computer development.

5.1.7 Strengths

Samsung enjoys the basic strength present in all other large Korean electronics corporations, diversity. In the electronics industry, with its pronounced business cycle swings, this is a particularly important attribute. Diversity aids the electronics sector in two ways: First, electronics production requires a diversity of skills, materials, and disciplines. For example, inclusion of a heavy construction business unit under a corporate umbrella may not help in day-to-day chip fabrication, but it certainly helps getting a billion dollar fab line in place on time and at a low cost. Second, diversity frees the corporation, as a whole, from the cycle of productivity and profit that is so significant in the semiconductor industry. When semiconductor sales lag, corporate profits from other sectors can sustain the base manufacturing capabilities. In fact, low-points of the semiconductor cycle can be thought of as "seed times" during which new products can be developed and new capabilities can be added by the work force. Samsung's new corporate management philosophy, with its emphasis on smaller, self-contained business units, might impede this "vertical integration."

5.1.8 Corporate Culture and Management Philosophy

Samsung calls itself an "employee-oriented" corporation. Its management practices are based on the belief that satisfied employees produce more for the company. The company is being segmented into smaller business units that appear to act as individual cost centers that act to "empower" individual employees. Promotion and monetary rewards are controlled from within a given business unit and are strongly tied to performance. Individual achievement is recognized and rewarded.

Emphasis on individual performance is not foreign to Korean industry. Company founders are accorded great status and prestige, both inside and outside the company, throughout Korea. But decentralization and emphasis on the performance of the single business unit might thwart the cooperative benefits of diversity described above. Other business units might not make allowances for the cyclic problems of SEC. It will be interesting to see how decentralization impacts Samsung's development over the next decade.

5.1.9 Vision and Strategy for Growth

SEC representatives did stress that, while they are fierce "competitors," they are not "predators." They feel that the company's success is based on its filling a vacuum in the world semiconductor market. The United States, for example, has concentrated on microprocessor development, and Japan alone cannot meet the world demand for DRAM parts. As such, Korean industry is symbiotic with, and essential to, the smooth functioning of the global semiconductor industry. The growth of Korean DRAM enterprise is dependent on the directions of the worldwide electronic systems market. If "memory hungry" systems (like HDTV or Internet graphics stations) prove to be the wave of the future, SEC will profit even more.

SEC is interested in careful diversification. For example, ASIC is regarded as a potential "vacuum" that the company can fill, at least in Asia. As the company's random logic design capability is somewhat weak at present, it is looking at universities as sources of expertise. Company managers routinely consult with professors; but more importantly, they are looking at the universities as "hot-houses" to grow the requisite talent.

In addition to the technological development plans described above, SEC is engaging in expansion of its physical plant capacity. This is facilitated by the large degree of vertical integration within Samsung Corporation as a whole.

5.2 LG SEMICON

LG Group, which was known as Lucky-Goldstar until 1995, adopted the new name to emphasize its evolution into a world-class company. The origin of the company goes back to 1947 when Rakhee Chemical Industry Company was founded. Today, the company consists of over thirty subsidiaries in the fields of chemicals, energy, electric, electronics, machinery, metals, trade, finance, construction and services, public services, and sports.

LG Group's involvement in the electrical and electronics industries, which started in 1959 with manufacture of a radio set, spans ten subsidiary companies (Fig. 5.1), one of which is LG Semicon Company, Ltd. LG Semicon was launched in 1983 by consolidating the semiconductor operations of two entities: Goldstar Electronics (founded in 1969) and Goldstar Semiconductors (founded in 1979). Figure 5.2 shows LG Semicon sales figures.

LG Semicon, which now employs over 7,500 people, has demonstrated phenomenal growth (Fig. 5.2). From 1990 to 1995, total sales shot up from U.S. $132 million ($70 million in exports) to $3.9 billion ($3.5 billion in exports, $1.3 billion in profits). Large production volume and an impressive series of technical accomplishments, as shown in Table 5.3, have made LG Semicon one of the major semiconductor companies of the world.

TRADE & FINANCE

LG International Corp.
LG Securities Co., Ltd.
LG Insurance Co., Ltd.
LG Merchant Banking Corp.
LG Credit Card Co., Ltd.

CONSTRUCTION & SERVICES

LG Construction Co., Ltd.
LG Engineering Co. Ltd.
LG Mart Co., Ltd.
LG Leisure Co., Ltd.
LG Ad Inc.
LG-EDS Systems Inc.
LG Economic Research Institute

PUBLIC SERVICE & SPORTS

LG Yonam Foundation
LG Yonam Educational Institute
LG Welfare Foundation
LG Sports, Ltd.

CHEMICALS & ENERGY

LG Chemical Ltd.
Honam Oil Refinery Co., Ltd.
LG Petrochemical Co., Ltd.
Hoyu Energy Co., Ltd.

ELECTRIC & ELECTRONICS

LG Electronics Inc.
LG Information & Communication, Ltd.
LG Software, Ltd.
LG Semicon Co., Ltd.
LG Electro-Components Ltd.
LG Precision Co., Ltd.
LG Industrial Systems Co., Ltd.
Goldstar Instrument & Electric Co., Ltd.
Goldstar Electric Machinery Co., Ltd.
LG-Honeywell Co., Ltd.

Figure 5.1. LG Subsidiaries

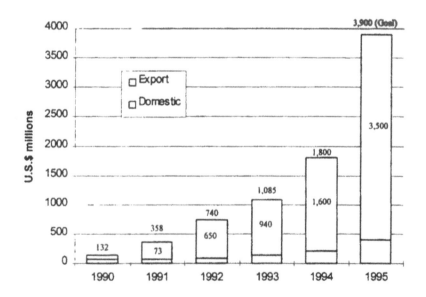

Figure 5.2. LG Semicon annual sales, 1990-1995.

Table 5.3
LG Semicon Accomplishments

Year	Milestone
1989	Goldstar Electron was founded, consolidating the semiconductor chip lines of Goldstar and GSS.
1990	Started operation of the Cheongju plant I.
1991	Started production of 4 Mb DRAM and developed 16 Mb DRAM.
1992	Developed and produced 64 Mb DRAM and 0.8 μm ASIC.
1993	Completed the construction of the Cheongju plant II, developed 16 Mb Mask ROM, and received ISO 9001 certification.
1994	Started mass production of 16 Mb DRAM, developed 32 Mb Mask ROM, and developed 0.6 μm ASIC.
1995	Changed company name to LG Semicon and started mass production of 64 Mb DRAM.

5.2.1 Organization

Unlike Hyundai, whose Electronics Industries, Ltd., subsidiary focuses not only on semiconductors but also on information systems, automotive electronics, and industrial electronics, LG Semicon focuses only on semiconductor components. It is divided into overlapping functional sectors: R&D, production, business, sales, and administrative support. Overlap exists between R&D and production sectors because of their coexistence in three major plants.

The three major LG Semicon plants are the Seoul plant (R&D on advanced memory and non-memory devices, design, and systems), the Cheongju plant (memories and R&D), and the Gumi Plant (memories, non-memory parts, and R&D). Of these, the Cheongju plant is the largest and the most advanced, having very high manufacturing yields. A new plant is soon to come on line at Kumi, Kyongsangpukto. According to company material, all the fabs are part of what is called the Technology R&D Unit.

Study of the organization of LG Semicon alone is difficult because it is well integrated with other LG subsidiaries (Fig. 5.1) in the electrical and electronics fields. In plant visits, it was somewhat difficult to distinguish between information presented about the LG Group as a whole and that of LG Semicon alone. LG Semicon employees think of themselves as employees of LG Group and seem to view everything, from goals and accomplishments to projects and products, with the perspective of the parent LG Group.

Distinguishing between subsidiaries is more difficult because multiple business units of LG Group are at the same site. For example, there is a concentration of company offices in Yongdunpo-gu district of Seoul. The site in the Seocho-Gu district of Seoul is the R&D headquarters of LG Semicon and is also the home for LG Electronics R&D group. The product display gallery, which contains a fascinating HDTV show, multimedia systems, computers networked for a video conference and, of course, microelectronics components; was obviously a joint effort of LG Semicon and LG Electronics. Thus, LG Semicon, though a dedicated semiconductor company, has allegiance to and extensive interactions with the parent company on the individual employee level and on the corporate level. The latter involves very strong strategic and tactical ties to other LG subsidiaries, which increasingly depend on LG Semicon for gaining prominence in non-component high-technology sectors. Based on the total sales figures, the internal business volume is currently very small, but an intense effort is underway to increase it.

5.2.2 Products and Technology

In 1990, Goldstar ranked 44th in semiconductor production and 15th in DRAM production in the world. In 1995, LG Semicon ranked 14th in world semiconductor production with 2% market share, and 5th in DRAM with 8.6% market share. LG is involved in all types of semiconductor technologies, many of which are used in electronics systems marketed by other LG subsidiaries. LG is making notable inroads into ASIC technology. LG's high-volume, flagship technology is a 0.5 μm CMOS process with three levels of metal. The first metal is tungsten with a TiN barrier; others are aluminum alloys. DRAMs are made in the leading technology with 0.5 μm metal/0.6 μm spaces. For ASICs, the metal width and spacing are both 0.7 μm. Cannon and Nikon lithography tools are used as work horses.

I-line steppers in conjunction with partial field phase-shift masks are used for 0.5 μm technology. Nikon steppers are used for 64 Mbit DRAMs. The company is switching to deep UV. R&D interests also exist in X-ray (using the Pohang light source), e-beam, and 193 nm lithography for 0.25 μm. The degree of LG Semicon involvement in X-ray lithography is unknown.

5.2.3 Research and Development

In 1994, LG Semicon spent 12% of its revenues on R&D. For 1995, the estimated figure was 8%. This R&D expense should be viewed in the light of LG Semicon's 1995 investment of $2.5 billion into new and refurbished facilities, an investment rate ranking next only to Intel and Motorola. The investment was expected to continue at a comparable level in 1996. R&D goals are two-fold: to maintain dominance in the memory market and to achieve dominance in non-memory applications by overcoming weaknesses in design, fabrication technology, and innovativeness, and by enlarging the captive market within the LG group.

The motivation and strategic guidelines for R&D seem to be derived from the LG Group as a single unit. The strong R&D effort is one element of the strategy for growth into what the LG Group's Chairman, Bon-Moo Koo, calls "a world-class business group in the 21st century." While being very proud of its manufacturing technology having the highest yields, the company is fully aware of this weakness: its world revolves around DRAM technology that came from Hitachi. The company realizes that its growth as a separate unit in the world market and its ability to support the Group's vision of becoming a global leader are critically dependent on developing its own technology base and making the transition from being a manufacturing company to being an innovating company. Hence, R&D activities are planned in four areas of the semiconductor field: circuit design, fabrication technology, CAD, and packaging. However, the R&D projects are clearly tied to products: 1 Gbit DRAM, 256 Mbit DRAM, 4 Mbit SRAM, 64 Mbit ROM, flash memory, PC peripherals, HDTV ICs, LCD ICs, Frame Memory, 16/8 bit MCU, 0.35 μm ASIC (standard ell), full custom ICs, and CCDs with 2 million pixels.

Extensive emphasis on R&D is also reflected in organization of LG Semicon into five R&D units based on products and long-term goals. The first R&D unit, called the ASIC & Memory Unit, focuses on DRAM, SRAM, and flash memories, and has been chartered to spearhead a thrust into ASICs (gate arrays, cell families, reusable ASICs,[1] and even CCDs) by leveraging the high-yielding memory fabrication technology. The second R&D unit deals with microcontrollers (currently 4- and 8-bit), smart card chips, and DSPs. The third unit, the Micro R&D unit, addresses all computer peripherals. The fourth unit, the Technology R&D unit, is the most important unit because it includes all fabrication facilities and product technology. A Systems & Devices subunit of the Technology R&D unit maintains the flagship fab process (currently 0.5 μm polysilicon gates, three-layer metal line/spaces 0.5/0.6 μm for DRAM and 0.7/0.7 μm for ASICs). A Design Automation subunit provides CAD tools for in-house use and for the world market. The fifth unit, Foreign R&D Labs (based in San José and London) gathers technology from around the world and develops information and media systems. Fifteen more R&D labs will be established by the year 2000.

5.2.4 Government and University Interactions

LG Semicon has been a participant in the Korean government's semiconductor memory research program. A quantitative evaluation of the benefits of the program to LG is difficult to make. LG engineers appear to share a common view that minimum government involvement in the affairs of industries will produce maximum results. It is interesting to note that LG representatives do not talk of doing away with R&D tax write-offs, but only about reducing government's regulatory authority.

The LG Semicon R&D unit has direct interactions with universities because it is located in the Seoul metropolitan area, which is home to a number of good schools. LG also has an X-ray lithography program in collaboration with Pohang University and light source but details were

---

[1] The term "reusable ASICs" is LG's creation; it simply means adding previously designed ASICs to the list of mega-cells (cores) available to customers.

unavailable. LG sponsors about 20 to 30 student projects with local schools, and it provides numerous (up to 400) student scholarships annually. LG engineers assert that the universities definitely do an important job of supplying excellent employees.

5.2.5 Alliances

The company, which came to prominence by implementing Hitachi's DRAM technology, continues to seek alliances for strengthening its infrastructure and seeking new technology and new products. The relationship with Hitachi continues with a joint venture on advanced memories. A license from Rambus (USA) will allow LG Semicon to produce a 16 Mbit synchronous RAM (ASIC memory). LG Semicon has collaborations with Sun Disk (USA) for flash memories, Compass (USA) on ASICs, and Siemens (Germany) on advanced microcontroller technology. It has established training relationships with U.S. universities, particularly Drexel for management training and the University of Illinois at Urbana Champaign for technical training. The importance of alliances is reinforced in the company's vision to "internationalize and globalize," to bring in technical resources from all over the world. The overseas research centers are chartered to track recent developments in technology and must be playing an important role in forming alliances and acquiring technology.

In a notable development in 1995, LG Semicon's sister business unit LG Electronics acquired a 57.7% stake in Zenith Electronics Corporation of the United States.

5.2.6 Overseas Activities

Most of LG Semicon's operations are located in the Republic of Korea, but the company does have overseas subsidiaries: San José, CA (an R&D center for information gathering and ASIC design house called Advanced MOS Technologies, Inc.), London (R&D), Hong Kong (two ASIC design houses, Valence Semiconductor, Ltd., and Syntheses Systems Design, Inc.). LG Semicon's overseas activities also include collaborations and alliances (some of which are discussed above). The parent group has been involved in technology transfer to foreign countries (Eastern Europe, the Middle East, Central and South America, and East Asia). LG Semicon has a worldwide network of sales offices. The parent group, which has a much wider presence on the global scene, provides the conduit for international business.

LG Semicon recently announced that it will build a state-of-the-art DRAM Fabrication facility in Malaysia. The project, in collaboration with Hitachi of Japan, will involve an investment of $1.3 billion and employ 1,000 professionals when started in early 1998. Initially the plant will produce 8.5 million 16 Mbit DRAMs per month and will also be capable of manufacturing 64 Mbit DRAMs.

5.2.7 Strengths
In addition to the previously discussed attributes that contribute to the strengths of Korean Companies, LG considers itself a pioneer of Korean Electronics Industry because it is the oldest electronics company (founded in 1958) to become an industrial giant. The structure of the parent group, in which three subsidiaries do business in electrical or electronic sectors, offers strong capital and captive market support while vigorously pursuing excellence in the risky business of semiconductor technology.

5.2.8 Company Culture and Management Philosophy
The corporate culture in LG Semicon is very similar to that in other companies in Korea and it revolves around the Corporate Group. There is a great emphasis on the corporate identity. All the engineers dress neatly but obviously according to a dress code and their jackets sport an LG logo. Fringe benefits such as company-provided housing, low-interest loans, and educational scholarships reinforce the corporate culture and enhance the bond between the employees and the company and also catalyze teamwork. Bonuses are used to enhance productivity. Other observations on the Korean corporate culture (see Chapter 1) obviously apply to LG Semicon.

5.2.9 Vision and Strategy for Growth
In broad terms, LG Group's vision is to become an international, global business leader that provides maximum value to the customer and nurtures a positive environment for the employees. According to a marketing and recruiting video, LG Electronics is and intends to remain *"a totally integrated electronics company."* This means globally dominating in everything and anything that concerns electronics. To accomplish this goal, LG Electronics' subsidiaries have chosen 20 electronics "items" or "fields" for future dominance. These choices represent either areas of explosive growth or technical fields that LG wants to dominate but which are currently dominated by other companies. HDTV, cellular communication, radars, robotics, and multimedia are examples from that list. All the electronics subsidiaries will target these fields for the purpose of gaining a global leadership position. The electrical and electronics companies (Fig. 5.1), in addition to having their own technology thrusts, have collaborative R&D and product development projects in these selected areas. LG Semicon, as a subsidiary of the LG Group, has a part to play in this vision. LG Semicon's Vice Chairman, Mr. Jung Hwan Moon, has launched the "Top Ten Spirit" campaign to become one of the world's top ten semiconductor companies by the year 2000 through acquisition of the highest-level chip-making technology, enhancement of quality and performance, and realization of maximum customer satisfaction. LG Semicon is also striving to be one of top five DRAM makers by the year 2000.

The nuts and bolts of this vision come from the middle managers who know the technologies and corresponding trends very well. These managers are also responsible for day-to-day management of employees with dignity and mutually providing a positive, nurturing environment needed to make the product. The vision, probably in terms of simple performance goals, is permeated to the employees in the trenches who participate in the vision by very hard work. LG's growth rate of 98.2% per year is a testimony to its tempo in this direction of its vision and the employee participation in it. If LG's track record, investment strategy, and alliance are taken as an indication of things to come, there is clearly substance to its vision.

5.3 HYUNDAI ELECTRONICS INDUSTRIES CO., LTD.

Americans are very familiar with Hyundai automobiles, which, since 1986, have been increasingly noticed by customers because of their low cost and steadily improving quality. Hyundai Motor Company is a subsidiary of the giant Hyundai Business Group, whose story is similar to that of the other successful industrial conglomerates. It started in 1947 as a humble small business, an automotive service shop. Today it has become an important player in the global market, with $76 billion in expected revenues. Its 180,000-person workforce has presence in over 30 countries. Hyundai claims to be Korea's largest company, as measured by workforce and revenues. It has over 76 affiliated companies all over the world, with businesses in a number of diverse areas (Fig. 5.3).

Founded in 1983, Hyundai Electronics Industries Co., Ltd., (HEI) is a major business unit of the Hyundai Business Group. Through a series of successes (Fig. 5.4 and Table 5.4), HEI has become one of the fastest growing electronics company in the world. Its $5.1 billion in 1995 revenues were expected to grow to $8.95 billion in 1996. It has more than 20,000 employees and assets totaling $5.4 billion.

5.3.1 Organization
Hyundai Electronics Industries has three major business sectors (Fig. 5.5), of which the semiconductor sector is the largest, having 76% of total 1994 revenues. The other two sectors, Information Systems (14% of revenues) and Industrial Electronics (10% of revenues), though smaller by revenue figures, are the fast-growing systems operations. A number of overseas operations that report directly to HEI's corporate offices are independently incorporated.

5.3.2 Products and Technology
Hyundai Electronics Industries' semiconductor product sector manufactures 1, 4, 16, and 64 Mbit DRAMs, has developed the world's first Synchronous 256 Mbit DRAM and is working on a 1 Gbit Synchronous DRAM. It also produces 1, 4, and 16 Mbit SRAMs. Nonmemory products include a variety of ASICs and a chipset for 586-based systems.

AUTOMOBILES

Hyundai Motor Company
Hyundai Motor Service Co., Ltd.
Kefico Corporation

IRON & METALS/ELECTRICAL EQUIPMENT/PRECISION

Inchon Iron & Steel Co., Ltd.
Hyundai Aluminum Industries Co. Ltd
Aluminum of Korea, Ltd.
Hyundai Pipe Co., Ltd.
Hyundai Elevator Co., Ltd.
Hyundai Precision & Industry Co., Ltd

TRADE & SERVICES

Hyundai Corporation
Hyundai Merchant Marine Co., Ltd.
Hyundai Wood Industries Co., Ltd.
Keum Kang Development Industrial
  Co., Ltd.
Hyundai Construction Equipment
  Service Co., Ltd.
Hyundai Securities Co., Ltd.
Hyundai Marine & Fire Insurance
  Co., Ltd.
KKBC International, Ltd.
Diamond AD, Ltd.

ELECTRONICS & INFORMATION SYSTEMS

Hyundai Electronics Industries Co. Ltd.
Hyundai Information Technology
Hyundai Tech System
Hyundai Media System

SHIPBUILDING INDUSTRIAL PLANTS

Hyundai Heavy Industries Co., Ltd.
Hyundai Mipa Dockyard Co., Ltd.

PETROCHEMICALS & OIL REFINERY

Hyundai Petrochemical Co., Ltd.
Hyundai Oil Refinery Co., Ltd.
Hyundai Tech System
Hyundai Media System

ENGINEERING & CONSTRUCTION

Hyundai Engineering & Construction
  Co., Ltd.
Hyundai Engineering Co., Ltd.
Hyundai Housing & Industrial
  Development Co., Ltd.
Koryeo Industrial Development Co., Ltd..

Figure 5.3. Hyundai subsidiaries.

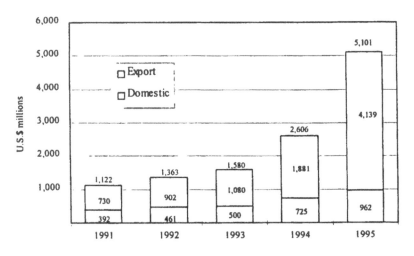

Figure 5.4. Hyundai Electronics Industries, annual sales, 1991-1995.

Table 5.4
Hyundai Accomplishments

Month	Year	Milestone
Feb.	1983	Established HEI
Mar.	1983	Founded subsidiary, HEA in United States
July	1985	Launched semiconductor assembly business
June	1988	Joined GSA program for PC in United States.
Nov.	1988	Founded subsidiary, HEE in Germany
Sept.	1989	Developed 4 Mbit DRAM
Mar.	1991	Developed 16 Mbit DRAM
Aug.	1991	Founded subsidiary, HES in Singapore
July	1992	Developed 64 Mbit DRAM
Jan.	1993	Participated in Digital Cellular System (CDMA) joint development project with ETRI
Feb.	1993	Developed domestic-first Car Navigation System (CNS)
July	1993	Developed HDTV working sample
Oct.	1993	Formed Strategic Alliance in Semiconductors with Fujitsu
Dec.	1993	Developed 64 Mbit DRAM engineering sample
Jan.	1994	Acquired partial ownership of MAXTOR, a manufacturer of HDD in United States
Mar.	1994	Participated in Globalstar Satellite Project
Apr.	1994	Founded workstation subsidiary, AXIL in United States
May	1994	Founded semiconductor assembly factory, HEC in China
May	1994	Participated in VOD (Video-On-Demand) business
Feb.	1995	Founded U.S. non-memory subsidiary, Symbios Logic Co.
Mar.	1995	Founded subsidiary, HEH in Hong Kong
July	1995	Founded subsidiary, HEU in U.K.
Sept.	1995	Developed 256 Mbit synchronous DRAM
Feb.	1996	Started to construct the memory fab in Eugene, OR (U.S.)

HEI has seven fabrication lines that use 5-, 6-, and 8-inch wafers and are capable of running processes with feature sizes 0.8, 0.65, 0.5, 0.4, and 0.35 μm. Although a 0.65 μm ASIC process is used most widely, DRAMs use a smaller feature size.

HEI ASIC technology is capable of providing 564,000 usable gates in conjunction with two-level metalization and 904,000 usable gates with three-level metalization technology. HEI has also produced a 32-bit, SPARC-compatible processor called Thunder 1.5. Although Hyundai engineers were aware of the SIA technology roadmap, they did not divulge any Hyundai technology roadmap.

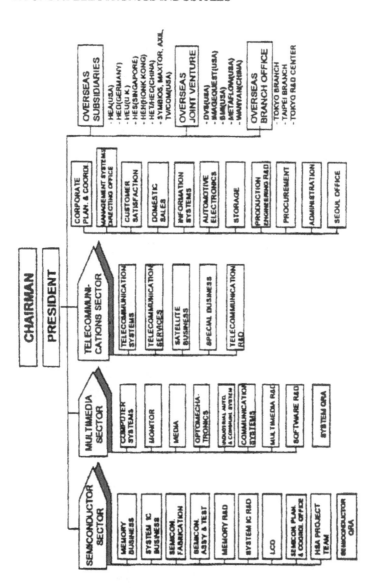

Figure 5.5. Hyundai's corporate structure.

HEI does all of its own packaging and does not subcontract work to packaging foundries. A large variety of SMD (surface mount devices), BGAs, PDIPs, hybrids, and memory cards are produced. HEI's 256 Mbit DRAM uses a lead on chip (LOC) packaging technology. Bond pads are interior to the chip, and hence, bond wires extend from the package post to the interior of the chip. An advanced packaging technology is implemented in the Smart Card, which contains a thinned memory chip mounted in a credit-card-sized unit. HEI's employee badges are all made of Smart Cards. Hyundai is in the process of developing chip-scale packages and considers the technology very valuable for future product lines.

HEI is heavily invested in passive and active matrix display devices. In late 1995, the latest technology was implemented in production of a 10.4 in. thin-film transistor (TFT)-based active matrix color display.

The product list for the information and industrial sectors of HEI is much longer than that of the semiconductor sector. These sectors produce and market computer, communications, TV, audio, video, network, multimedia, copying, photographic, and automotive electronics products. On the low end of the technology is a copying machine based on technology licensed from Canon, Japan. The high-end example of HEI's technology is Korea's first Satellite called "PagerSat," which will accommodate up to 10 million pager subscribers. The satellite, developed jointly by HEI's information systems sector and the Electronics and Telecommunications Research Institute (ETRI), was to be launched in 1996. Link and reception technology and ground stations (up to 240) have already been tested. HEI has a complete line of automotive electronics systems and products that it sells to all the major auto makers in the world. The latest automotive product is a navigation system based on GPS (Global Positioning System).

5.3.3 Research and Development

Hyundai Business Group operates four subsidiaries in the electronics and related high-technology areas: Hyundai Electronics Industries (HEI), Hyundai Information Technology, Hyundai Tech Systems, and Hyundai Media Systems. The electronics and component business is consolidated. As a single company spearheading a campaign to become one of the top five electronics companies in the world by the year 2000, HEI can orchestrate coherent R&D efforts in all three areas of its emphasis: semiconductors, industrial electronics, and information systems. HEI's public relations brochure describes its R&D activity as awesome. HEI channels 12% of its revenues into R&D carried out in five laboratories. The overall focus of all the R&D is on marketable products. Two laboratories are devoted to semiconductor products, one laboratory to industrial product development, and two laboratories to information systems R&D.

Two semiconductor R&D laboratories, divided between memory and nonmemory product lines, employ about 1,000 professionals who are extending the technology using commercially available materials and tools to make next-generation products. The *Memory R&D lab*, is developing a 1Gbit DRAM and also focusing on the manufacturability of 64 and 256 Mbit DRAMs, synchronous DRAM, synchronous graphic DRAM, low-power 16 Mbit SRAM, 16 Mbit flash memory and 32 Mbit mask ROM. Hyundai representatives were particularly proud of the recently marketed 4 Mbit DRAM and made it a point to mention that it is a quality product with the best specifications. The *System IC R&D lab*, focuses on non-memory chip development: RISC/CISC processors, PC chipsets, cellular phone chipsets, multimedia and video chips, and 0.5 micron ASICs. Hyundai hasn't had the time or the resources to engage in the advanced solid state research, which was once the staple for leading U.S. R&D labs.

The multimedia sector's software and multimedia R&D laboratories concentrate on multimedia and high-speed network technology. Software R&D is aimed at designing systems and networks for the future. The Software R&D lab played an important role in the development of the SPARC processor in collaboration with METAFLOW of the United States. The multimedia R&D lab works on advanced building blocks of multimedia systems (e.g., blue laser and MPEG chips) and on products such as personal computers, video CD players, LCD projectors, game machines, HDTV, interactive cable TV, and video on demand.

The telecommunication sector's R&D activities are focused on core technology for next-generation products. Core areas of work are mechanisms (e.g., high-precision mechanisms in cameras, copiers, disk players, remote control devices, etc.); communications including mobile systems; computer aided engineering (CAE) and ASICs; equipment and instruments; and automotive electronics. The Industrial R&D lab developed an automotive navigation systems based on GPS. It was a partner with ETRI for the development of the PagerSat.

5.3.4 Government and University Interactions

Hyundai's perception of the role of the government and universities is very similar to that of the other semiconductor companies; however, a few comments specific to Hyundai are made here.

Whatever company executives' perception of the government's accomplishments may be, according to a recent news report, Hyundai has clearly benefited from government involvement in R&D. The PagerSat project that Hyundai developed jointly with the government-funded R&D lab ETRI is now near completion and due to be operational in 1996.

HEI has been a contributing partner in a government-sponsored semiconductor memory R&D program. However, the extent of the participation could not be determined. From Hyundai management's point of view, the government's action, or the lack of it, and excess regulation have hurt it economically on specific occasions. Similar to other companies, executives at Hyundai obviously would like to see less government involvement in industrial activities.

Hyundai does all its own research.   Since there are no large universities near the HEI site, the only interactions it has with universities is through a large number of scholarships, donations (such as the one that helped build a new design center at the Korean Institute of Technology), and employee training at the universities.

5.3.5 Alliances

Hyundai Electronics Industries has formed numerous technical alliances and invested in small companies to obtain critical technologies. Since October 1993, it has had a technical cooperation pact with Fujitsu of Japan on DRAM development.   In the breakup of NCR (National Cash Register Corporation), which was acquired by AT&T, Hyundai purchased NCR's non-memory semiconductor business from the new owner for $340 million and named it Symbios Logic, Inc.   This acquisition provided HEI with a proven ASIC technology.

In 1994, Hyundai invested $391 million to buy Maxitor Corporation, a U.S. disk drive maker that expects sales of $5 billion in the second half of the 1990s. This relatively small company is trying to make a comeback as a subsidiary of Hyundai Electronics America. In January 1996, it received a bridge loan of $100 million to provide working capital for three months.

HEI has a small collaboration with Laserbyte Corporation of Sunnyvale, CA (USA) on the development of magneto-optical disk drives. HEI accomplished the development of a 32-bit, SPARC-compatible processor in a joint venture with Metaflow of La Jolla, CA, in which HEI provided Metaflow with approximately $1.7 million in capital.   Hyundai also has joint ventures with Imagequest of Fremont, CA (TFT-LCD development, $1 million), and BMI of Santa Clara, CA, (flash memory, $3 million).

5.3.6 Overseas Activities

HEI has a global presence in the form of over ten subsidiaries, R&D centers, and overseas branches. Other than the branch offices, which are marketing arms of HEI-Republic of Korea, all the overseas subsidiaries have broader functions in manufacturing, marketing, and sales of semiconductor components, computers, communications gear, car audio systems, direct broadcasting, and digital CATV. The R&D activities are in semiconductors and computers. Table 5.5 lists HEI's overseas subsidiaries.

5.3.7 Strengths

Attributes that strengthen Hyundai Electronics' position as a global leader in high technology products are similar to those of other Korean companies and are discussed in Chapter 1. The younger age (on average) of employees, their patriotism, dedication, hard work, culture, and tradition, manufacturing excellence, strong leadership, good capital support, R&D investment, corporate perseverance, and company-wide passion to become

Table 5.5
Hyundai Overseas Activities

	FOUNDED	LOCATION	CAPITAL
SALES SUBSIDIARIES			
• Marketing and sales of semiconductors, computers, monitors, and multimedia products • R&D activities for semiconductors and multimedia (HEA)			
Hyundai Electronics America (HEA)	March 1983	San Jose, CA., USA	$404 M
Hyundai Electronics U.K. (HEU)	July 1995	London, U.K.	$1 M
Hyundai Electronics Deutschland GmbH (HED)	November 1988	Frankfurt, Germany	$2.7 M
Hyundai Electronics Asia PTE, Ltd. (HES)	August 1991	Singapore	$0.3 M
Hyundai Electronics Hong Kong (HEH)	March 1995	Hong Kong	$1 M
PRODUCTION AND INDEPENDENT SUBSIDIARIES			
MAXITOR (HDD)	January 1994	San Jose, CA, USA	$391 M
Hyundai Electronics Tianjin Co., Ltd. (HET) (Car audio)	January 1994	Tianjin, PRC	$1.5 M
Axil Computer Inc. (AXIL) (Workstations)	April 1994	Santa Clara, CA, USA	$30 M
Hyundai Electronics Shanghai Co., Ltd. (HEC) (Semiconductor Assy & Test)	May 1994	Shanghai, PRC	$33 M
Symbios Logic, Inc. (SLI) (ASICs)	February 1995	Ft. Collins, CO, USA	$340 M
TV/COM (Digital CATV & DBS)	July 1995	San Diego, CA, USA	$22 M

Source: Hyundai Electronics Industries Co., Ltd.

number one are all factors that make Hyundai a strong contender in the race for global leadership. Hyundai, in particular, is financially stronger because it is a very large, diversified conglomerate.

5.3.8 Corporate and Management Philosophy

Hyundai culture and management philosophy are very similar to those discussed before that revolve around employees. Hyundai seems to enjoy a notably harmonious employee-management relationship. According to the panel's hosts at Hyundai, the employee unions in Hyundai are very friendly. The company's philosophy, modified from an old Korean saying, "Body and soul are not two" (i.e., they are an inseparable team; separation will mean death), is "Management and labor are not two." A logo advertising this campaign company-wide on posters and on buttons worn by employees is in the Korean script and looks like two stick figures holding hands, symbolizing cooperation between management and workers. The symbolism of equality between management and engineer workers is reflected in the uniform they wear. The engineers and the Managing Director who hosted the JTEC/WTEC team all wore white shirts, navy blue trousers and dark blue jackets displaying both the company logo and the teamwork logo.

The company emphasizes teamwork and identity with the company. The managing director of the memory lab, Mr. Soo Han Choi, commented that Korea doesn't have any individual star players. Everything is a team effort. He was convinced that future breakthroughs in the semiconductor field are going to come out of teamwork. Then, after comparing an evolution to a revolution, he went on to argue that innovation, according to Hyundai, was going to come about as an aggregate of small advances. Innovative and novel products will be made in the future on the basis of "cumulative efforts put in by a large number of individuals."

Frugality and modesty are also a part of Hyundai's corporate culture. The Hyundai Group and HEI have intentionally avoided propaganda or public relations campaigns that do not impart higher quality or value to products but simply enhance the image. Hence, Hyundai representatives assert, the quality and value of Hyundai products are always better than their image. They cite the examples of Hyundai automobiles sold in the United States and also the recently introduced 4 Mbit DRAM. (Incidentally, Hyundai USA ran two-page advertisements about its DRAM starting in November 1995.)

On the relationship between quality, value, and cost, Hyundai representatives assert that people don't buy products only on the basis of an absolute standard of quality. They decide to buy a product using a norm that defines its value to them, an acceptable (or tolerable) level of quality, and the money (cost) they are willing to pay for it. Hyundai products are designed and manufactured to maximize customer acceptance.

5.3.9 Vision and Strategies for Growth

Hyundai Electronics Industries intends to achieve "ultra-super enterprise" status by the turn of the century by emphasizing multimedia and focusing on all associated technologies: information processing, storage, and communication. The company video proclaims the strategy for growth is globalization without ignoring localization. Other components of the growth strategy are constant innovation, implementation of advanced technology, and what Hyundai calls "humanization." Hyundai's products, projects, corporate philosophy, and business plans reflect this strategy for growth. For example, Hyundai's Oregon semiconductor plant will be the largest in the United States and will allow Hyundai to respond to the local market and political needs while taking Hyundai manufacturing capabilities global. The $1.8 billion Global Star project will provide worldwide communication coverage via Hyundai satellites and use Hyundai electronics equipment. The PagerSat project, which is nearing full operation, will allow global paging service.

HEI's Corporate Product Showroom reflects the Hyundai strategy: One section displays a complete set of automotive electronics subsystems, all of which are used in Hyundai automobiles and are also sold to other car manufacturers. The list of automotive electronics products that are in the product development cycle includes the global positioning system, collision avoidance unit, and advanced driver information center. While showing a copier, the showroom guide pointed out that the copier technology, though purchased from the Japanese, reflects Hyundai's manufacturing capabilities and makes Hyundai's information processing product line complete. Examples of product innovation are a remote control camera and an all-in-one home security center that includes a phone, a display for a remotely located video camera(s), a paging or door-answering system, a remote control door unlocking system and a fire alarm control center. Implementation of advanced technology is most evident in the semiconductor component sector.

A more practical goal for employees is described in a competitive spirit: Hyundai wants to be one of the top five semiconductor companies by the end of the century. Hyundai staff believe that of the five top companies in the year 2000, up to three will be Japanese and at least two will be Korean! This vision, probably in terms of simple performance goals, permeates through the company to the employees in the trenches who participate in the vision by very hard work. The middle managers are responsible for the day-to-day management of employees and for maximizing productivity without dehumanizing the organization. Hyundai's growth rate of 98% per year is a testimony to its tempo in the direction of its vision and employee participation in it. If Hyundai's track record, investment strategy, and alliances are taken as an indication of things to come, there is clearly substance to its vision.

5.4 DAEWOO GROUP

Daewoo Group is a conglomerate of 30 firms in many different areas, providing services ranging from automotive to financial. The Daewoo companies Daewoo Electronics and Daewoo Semiconductors and their divisions and subdivisions are relevant to this study.

5.4.1 Daewoo Electronics

Daewoo Electronics Company was founded in 1974 as one of the major members of the Daewoo Group. In 1995, Daewoo Electronics expected total sales of about $4.35 billion, a 35% increase over 1994 sales. The figure includes over $2.8 billion in exports, a 47.4% increase over the previous year.

Daewoo Electronics is divided into six divisions: computers, audio, video, home appliances, components, and systems. The six divisions are described briefly below. The company's Korean production facilities are shown in Table 5.6.

- *Computers Business Division (CBD).* The division's main customers are IBM, Amstrad, Commodore, Tandon, Headstart, and Cordata Technologies. CBD has been able to offer advanced computers including the Pentium and 486 Desk/Laptop computers.

- *Audio Division.* The audio division in Daewoo Electronics controls OEM business for international brands such as Toshiba, Sony, and Marantz.

- *Video Division.* The video division produces approximately 2 million VCRs, 27 million color TVs, 8 million black & white TVs and 200,000 camcorders each year. The division is also developing, upgrading, and improving products such as HDTVs, LCD TVs, Teletext TVs, Digital TVs, Monitors, Digital Hi-Fi Stereo VCRs, monitors combined with video cassette recorders, 4-Head VCRs, SVHS, and 8 mm Camcorders.

- *Home Appliances Division.* The home appliances division produces refrigerators, washing machines, microwave ovens, vacuum cleaners, air conditioners, and other home appliances.

Table 5.6
Daewoo Electronics, Domestic Production Facilities

Facilities	Area size (m²)	Products
Kumi Complex	734,000	B/W TVs, C/TVs, VCRs, camcorders, monitors, computers
Kwangju Complex	357,000	Audio systems, microwaves, washing machines, air conditioners, vacuum cleaners
Ichon Plant	119,000	Refrigerators, fans, heaters, compressors
Juan Plant	15,000	VCR decks
Jungju Plant	23,500	Deflection yokes, RF modulators, tantalum capacitors, hybrid ICs

- *Daewoo Electronic Components Division (DECD).* This division manufactures electronic components for home appliances and consumer electronics, including handphone power supplies, deflection yokes and flyback transformers for TV and home appliances, and aluminum electrolytic and tantalum capacitors. The company expected sales for 1995 to total more than $231 million, including $193 million in exports. Production facilities are overseas as well as in Korea.
- *Systems Division.* The systems division provides consultation, design, installation, maintenance, and special requirement services for electronic systems. The typical system includes integrated engineering in audio/visual, security, data communications, telecommunications, office automation, and building automation.

Daewoo Electronics also has three subsidiaries that are involved in electronics, briefly described below.

Orion Electric Co., Ltd., was established in 1965. Orion Electric had estimated sales of $979 million for 1995. Color picture tubes (CPT) and color display tubes (CDT) accounted for about 85% of the company's exports. Asia took 52% of Orion Electric's exports, with 27% going to the Americas, and the remaining 21% going to Europe.

Orion Electric exists as three plants located in the Gumi Industrial Complex south of Seoul. Its main productions are B/W television tubes, color picture tubes, monitors, electron gun mount parts, oscilloscope CRTs, and other products. The company has developed high-resolution display tubes, minideck CRTs, and large CRTs by employing the latest research results into design, and making excellent use of fully automated production systems. About 90% of Orion's sales are from exports, which exceeded $270 million in 1989. The company is gradually growing its international market while enlarging its production section to around 4,000 employees.

Daewoo Carrier Corporation was established in 1985 with the U.S.-based Carrier Corporation. The company manufactures heating and cooling systems, in addition to other related products in the Southeastern Hanam Industrial Estate facility. This modern facility has an annual production capacity of 800,000 rotary packaged air conditioners and related products. They also produced 5,000 to 24,000 BUTH high-energy efficient rotary compressor. Other main products are furnaces, fan coil units, condensing units, and cooling towers.

Daewoo Telecom Co., Ltd., developed a fully time division switching system (TDX) that helped move Korea into the number ten spot in terms of market share in telecommunication systems. This company has been developing public and private switching systems, optical transmission systems, radio communication systems, and broadcasting systems. It also produces computer systems, including the Pentium PC, engineering workstations, compact laptop computers, and super mini computers.

Daewoo Electronics' divisions have substantial overseas production, R&D, and sales interests. In sum, Daewoo Electronics divisions increased their number of overseas sales subsidiaries from 18 to 21 during 1995, with new ones in Bangkok, Lima, and Bucharest. The company now has ten sales subsidiaries in Europe, seven in the Americas, one in Asia, two in Russia, and one in the Middle East.

Also in 1995, Daewoo Electronics expanded its number of overseas production plants from 16 to 19. Offshore production plants number five in Europe, four in China, six in the rest of Asia (including Malaysia and Vietnam), two in the Americas, and two in the former Soviet Union. These plants produce a variety of consumer electronics products, including large and small home appliances, televisions, and car stereos. Annual overseas production capacity of color televisions by Daewoo Electronics amounts to over 50% of all its color TV production capacity. (The company expects to capture 12% of the global television market by the year 2000 with sales of 12 million units.) The company is also rapidly increasing the ratio of overseas production in other major home appliance products (VCRs, microwave ovens, washing machines, and refrigerators), and it already produces some 57% of its car stereos abroad. Daewoo's — and the world's — largest facility for the production of VCR heads (up to 10 million annually, with a technical license agreement from Toshiba) is in Hanoi, Vietnam. A huge new regional base near Warsaw, Poland, when completed, will produce large home appliances, consumer electronics, and related components in addition to the color TVs already produced there. Overseas production in all products reached 35% of total production in 1995 and the company expects this figure to climb to 60% by the year 2000.

Daewoo Electronics is developing a worldwide network of research centers and opened its first overseas integrated research center in France in mid-1995. This is the company's fourth overseas research center, following one in Japan and two in France. By 2000, Daewoo Electronics expects to establish seven additional design and research centers overseas in the United States, Mexico, Japan, China, Northern Ireland, Poland, and Russia. These centers are developing products to meet regional design and performance requirements, as well as coordinate activities to develop products that offer worldwide appeal.

Daewoo Electronics' export agreements in 1995 include the following:

- a $65 million deal with Benha Company for Electronics Industries of Egypt for export of a color television plant and 450,000 color television kits for assembly in Egypt
- agreement to supply Talcar Corp., Ltd., of Israel $50 million worth of Daewoo brand products through 1997
- a $25 million order from the China National Industrial Machinery Import & Export Company to provide systems and equipment for a new airport being build in Jilin Province, China

Daewoo Electronics is headed towards a goal of zero defects in production and established its own Quality Management Research Center in Seoul. The company has ISO 9001 certification for virtually all of its processes, products, and programs, solidifying its reputation around the world for full quality. Also as a result of fortified production quality control, Daewoo Electronics received two Japanese marks of quality approval: the "S-JQA" mark for its color televisions and TVCRs, and the "Good Design" mark for its VCRs and microwave ovens. Daewoo Electronics also received Canadian CSA approval for its core products.

5.4.2 Daewoo Semiconductors

The Daewoo Group came late to the Korean semiconductor industry. Daewoo Semiconductor was established in 1984 through 50% collaboration with Northern Telecom. In 1986, the Daewoo-Zymos Technology Company was formed through a 51% collaboration with the U.S. semi-custom semiconductor firm Zymos. In the same year, Daewoo signed a 1M Mask ROM export contract with the Japanese company Ricoh, and went into the Guro Industries estate area (on the outskirts of Seoul) and transformed the garment and wig area into a fab, importing equipment from Zymos. The Seoul location was chosen primarily because of proximity to universities.

The Daewoo Group adopted a niche strategy that avoided the standard memory division in which Samsung and Hyundai had invested heavily; it instead concentrated on a small-scale customized semiconductor division. Initially Daewoo looked to Zymos for design and manufacturing, but later it transferred the design to Korea.

Daewoo Semiconductor employs about 420 people, including 15 bipolar design engineers, 45 MOS design engineers, and 30 process engineers. Revenues were $58 million in 1995, with expectations they would increase to $78 million in 1996. Key products include consumer audio, video and telecom devices, 4- and 8-bit MCUs, gate arrays, and standard cells. The company has 2 fab lines with 4-in. wafer capabilities. It also has an assembly line for SIP, ZIP, and heat sinks, and both digital and analog test facilities. The process technology includes standard and high-frequency bipolar (fT: 6 GHz) and 1.2 μm CMOS.

Initially, Daewoo Semiconductor sold primarily in the domestic market, but gradually sales have widened. Today only about 50% of products are for internal consumption. The Daewoo Group provides an internal market in which products can be field tested before being externally marketed.

Daewoo intends to expand into some key semiconductor technologies. It currently plans to develop a new semiconductor fab and is considering technology and infrastructure needs. It is clearly looking outside Korea for a possible location. The Daewoo Group construction company would build the fab using Korean fab designers. Equipment would be purchased from Japan and the United States, and back-end assembly and testing could be done in Korea or outside Korea using Korean equipment.

5.5 ANAM GROUP

Anam is a group of businesses that is primarily involved in semiconductor packaging but in 1986 branched into construction and has plans to develop a broader semiconductor device fabrication base, largely in the ASIC arena. Nine companies comprise the Anam group:

1. Anam Industrial Co., Ltd.

2. Anam Electronics Co., Ltd.

3. Anam Engineering & Construction Co., Ltd.

4. Anam Instruments Co., Ltd.

5. Anam Semiconductor & Technology Co., Ltd.

6. Korea National Electric Co., Ltd.

7. Amkor Electronics, Inc.

8. Amkor/Anam Philipinas, Inc.

9. Automated Micro Electronics, Inc.

The first six companies are headquartered in Korea; Amkor Electronics, Inc., is headquartered in the United States; and the last two companies are headquartered in the Philippines.

The original company, Anam Industrial Co., Ltd., was established by Hyang-Soo Kim, the father of the current chairman, in 1956. Anam Industrial and Anam Electronics are publicly held companies on the Korean Stock Exchange. Amkor Electronics, Inc., is privately held by Joo-Jin Kim, the current Chairman of the Anam Group. Amkor/Anam Philipinas, Inc., is jointly owned by Amkor Electronics (60%) and Anam Industrial (40%). Automated Micro Electronics, Inc., is wholly owned by Amkor/Anam Philipinas.

Anam Industrial is made up of three divisions, one of which is the Semiconductor Business Division, which is the semiconductor assembly and test business. Anam Semiconductor & Technology Co., Ltd., is focused on ASIC, semiconductor equipment and parts, and wireless LAN. Amkor Electronics is the global sales and manufacturing arm of Anam Industrial. Amkor facilities are identified as Anam/Amkor, which often causes confusion. The Quality and Reliability Engineering Division vice president of Amkor resides in Korea to provide leadership in this arena to Anam Industrial. Anam/Amkor maintains a 30-40% market share of the microelectronics subcontractor packaging business and has the majority of major semiconductor manufacturers as its customers. The top competitors are Hitachi in Japan, Alphatec in Thailand, and Hyundai in Korea. Anam/Amkor has approximately 6000 employees.

5.5.1 Facilities

Having invested $500 million in packaging in 1995, Anam Industrial plans to exceed $4 billion in sales in the year 2000. Dr. Hwang believes that the company is well positioned to maintain its lead, due to its 28-year experience in the business and the opportunity to accumulate technology over this long period of time. He recognizes the large capital investment needs of growing in this business and says the financial success of the company enables the necessary outlays. To meet the growing demand for production in Korea and continue with the vertical expansion in semiconductor technology, ground was broken for a new plant at Kwangju City in October of 1995. This is the group's fourth manufacturing facility in Korea, and it projects it will have five buildings for assembly by 1999. Anam's first silicon fabrication capability is planned to be completed by 2000. The current plan is for providing 0.35 μm technology for ASIC companies. With the ASIC capability of Anam Semiconductor & Technology Co., Ltd., Anam will be positioned to become a major player in the ASIC market.

Kwangju City, a high-tech science industrial development, is 250 miles south of Seoul, and the Anam site comprises 105.6 acres. Anam/Amkor is expanding into Minsk, Republic of Belarus, in a joint venture with Integral and Krause, to serve the European market. The Anam group currently has three packaging facilities around the greater Seoul area and three in the Philippines. The ball grid array is planned to be manufactured in Chandler, Arizona, in the United States. The corporate headquarters of Amkor are located in West Chester, PA, with sales and support offices in Singapore, France, Taiwan, Tokyo, Santa Clara, and Dallas, in addition to the group's manufacturing sites. Design centers are located in Dallas and Chandler.

5.5.2 Technology Capability

Anam has a history of importing technology and then building upon it. Early collaboration with Japanese companies led to the building of the first Korean television. Matsushita provided components and DuPont provided photomask technology. More recently, Anam has collaborated with VLSI, Inc., to develop ASIC capability. Power-dissipating packages were introduced in 1992 and ball grid array (BGA) packages were available in 1994. Activities slated for 1996 introduction were multichip modules, C-4 BGA, and electrically enhanced (reduced inductance) packages. The C-4 activity is multipronged, with both external and internal approaches to design, materials, and construction being considered. Amkor has been an active member of the standards committees on package reliability and form factor.

Anam actively uses general industry roadmaps and also works with customers to develop its own strategy. Anam management and engineers meet with NEC and Toshiba four times a year to exchange views on industry trends. This has resulted in an exchange of R&D engineers for closer relationships. Three times a year, Anam updates its own technology roadmap, which is shared freely with its customers. The philosophy is to change as little of the manufacturing process as possible with each new package introduction, thereby leveraging the experience and knowledge to introduce a high-quality, high-reliability product from the outset. Anam's wirebonding capability is at 0.3 mm, which exceeds the current requirements of U.S. and Japanese board manufacturers.

Other Anam companies develop much of Anam Industries' manufacturing equipment. The trim and form and molding equipment is not offered to other companies and often is supplemented with purchased equipment from Japan and the United States. The original development and technological expertise was based in the United States and moved to Asia over the last ten years. With the introduction of new packaging concepts, there is a trend to see this expertise returning to the United States, but with strong Asian company participation.

Currently, testing of packaged products is only about 10% of Anam Industrial's business portfolio. This is projected to increase as many companies strive to reduce their overall cycle time. Anam is seeing an increase in package, test, and drop-ship requests from fabless design houses. This turn-key solution includes burn-in, bake, and dry pack. Anam does not provide programming services for PALS or Flash devices.

5.5.3 Quality and Reliability

Anam is ISO 9002- and QS 9000-registered. It has a data system with customer access to its extensive reliability databases. Amkor is developing a multilevel qualification methodology that will allow customers to utilize products at various stages of maturity with a clearly articulated and assessed risk. A joint project with a U.S. customer is currently underway to develop and evaluate an electrical and thermal reliability model.

Internally, the company's focus is to empower the workforce to make decisions at a lower level and increase productivity. Operators have been trained to do maintenance on the line equipment, and quality circles abound. The visible enthusiasm of the Korean culture has allowed many areas to progress quickly, although with some setbacks along the way. When provided with clear objectives and training, the workforce moves as one with significant momentum

5.5.4 University and Government Interactions

Anam works with universities and institutions through a consortium with the Korean government. Funding for these programs are shared equally between the company and government. Since 1993, a good response has been received from these projects, with the highest success rate at Seoul University. Anam follows worldwide academic publications and utilizes those works in its operations.

5.6 SIGNETICS KOREA COMPANY

Signetics Korea was established in 1966 as part of Signetics Corporation. In 1975 it joined the Philips Corporation. In 1995, ownership of Signetics Korea was transferred to Keo Pyung, which is a diverse, Korean-based group that has 15 major businesses, primarily involved in construction and retailing.

5.6.1 Business Brief

Signetics Korea is a subcontract house for packaging and testing of integrated circuits. Current packages include PDIP, SOP, QSOP, SSOP, and PLCC. Its business focus has successfully been on world-class quality assembly and customer satisfaction for just a few package types. The company now intends to build on its success by progressively adding more package types. Starting in September 1996, a new $75 million plant near Seoul (Kyunggi-do) will be utilized to create assembly lines for QFP, TQFP, SSOP, TSOP, TSSOP, SOJ, BGA, and CSP package types.

5.6.2 Major Customers and Competitors

Major customers of Signetics Korea include HP, IBM, NCR, Digital Equipment, DELCO, Ford, Chrysler, AT&T, Wang, NEC, Matsushita, Seagate, Atari, Atmel, Apple, General Electric, Unisys, Valence, Xerox, Intel, SGS-Thompson, Sharp, Samsung, Goldstar, Daewoo, SCI U.K., and Synergy. Its major competitors are Anam/Amkor and ASE.

5.6.3 Reliability and Failure Analysis

Signetics Korea has a self-qualification and reliability program to monitor the device integrity. Reliability testing includes data collection to detect product anomalies and monitor trends. Failure analysis facilities are available to support test-related issues for further improvement and customer satisfaction. Major failure analysis equipment includes SEM (ISI SX-40 and ISI Super-IIIA), optical microscope (Nikon UFX-II), plasma etcher (Technics RIE800), X-ray (Softex), EDAX (Philips PV9900), and probe station (Signatone).

5.6.4 Quality

The quality policy of Signetics is to manufacture and deliver defect-free and reliable products with on-time-to-request dates and the lowest possible cost by doing the job right the first time. They have a Total Quality Management program that has been certified by ISO 9001, SAC level 1, QS 9000 and EIA599. Operators are 100% quality trained, have 100% participation in quality teams, and have approximately 20% turnover per year. They have a near 99% on-time delivery history over the last two years, except during a few summer months where the averages dropped due to employee vacation. The engineering group has an average of 7 years of experience and is comprised of approximately 40% electrical engineers, 40% mechanical engineers, and 20% engineers with a variety of background in physics, chemical engineering, materials science, etc. Automated visual inspection is performed on all packages and has yielded 54 PPM or less over the last three years. Human interaction is viewed as the biggest barrier to further quality improvement.

5.6.5 Final Test House

One of the very strong points of Signetics Korea is its electrical testing capability. Full electrical testing is available for logic, linear, memory, PLD, FPGA, communication, MCU, and controller devices. Dynamic and static burn-in testing is available, as well as three-temperature testing. Major test equipment includes Accu-test (ACCU7702), Advan-test (T3340, T3313, and T3319), Teradyne (A311, J2773, J283, J387, A360, J941, and J983), Megatest (Q2/52 and Q2/62), MCT (2020 and 2010), Schlumberger (Sentry 10/21), and Credence (VI1001, VV1001, and STS350). Quality testing for electrical parameters is approaching 0 PPM.

5.7 SUMMARY

The Korean "giants" of electronics began operations in areas such as agriculture, construction, and chemical supply. As market trends changed, they diversified into electronics, making use of technology transfer from Japan and the United States and personnel trained in the electronics industry.

Korean firms are now leaders in many areas of high-technology device and process development, most notably the DRAM market segment. The next ten years look very good for the Korean electronics companies.

Chapter 6

KEY INSTITUTES AND UNIVERSITIES
SUPPORTING ELECTRONICS

This chapter profiles some of Korea's important educational institutions that contribute both to training of next-generation engineers and technicians and to performing basic and applied R&D that support the country's electronics industry

6.1. ELECTRONICS AND TELECOMMUNICATIONS RESEARCH INSTITUTE (ETRI)

The Electronics and Telecommunications Research Institute (ETRI) has evolved since 1976 through a series of predecessors and has been affiliated at different times with the Ministry of Trade, Industry, and Energy (MOTIE), what is now the Ministry of Information and Communications (MOIC), and the Ministry of Science and Technology (MOST). In 1992, ETRI became an affiliated research institute of MOIC, and in 1995, ETRI became an affiliated corporate body of MOIC.

6.1.1 Major R&D Areas
The following describes a few of ETRI's many important technology breakthroughs:

- In the telecommunications network area, in order to construct new social overhead capital, ETRI intensively drives forward the development of the ultra-high-speed information technology (IT) network, the Advanced Intelligent Network (AIN), and Universal Personal Telecommunications (UPT).

- In the switching area, ETRI has developed ATM switching, mobile switching, intelligent switching, and optical switching, and has begun study of next-generation switching technologies.

- In the transmission area, ETRI is proceeding with R&D to develop the 10-Gbit/100-Gbit transmission systems, BDCS, and B-NT as a core technology of B-ISDN.

- In the mobile telecommunications area, ETRI is intensively driving forward the development of CDMA, PCS, mobile data telecommunications, mobile satellite telecommunications, and radio wave resources utilization for the 21st century.

- In the satellite communications area, ETRI is proceeding with R&D to develop not only the earth station, TT&C, and payload subsystem for Korea's *Mugunghwa* satellite, but also mobile satellite telecommunications and digital broadcasting satellite telecommunications with broadcasting.

- In the computer area, ETRI also focuses on developing high-speed parallel processing computers and intelligent computers as well as distributed system software, data service technology, and human-computer interface technology.

- In the semiconductor area, ETRI is intensively driving forward the technology development of IC design for IT, high-speed and optical devices, advanced Si devices, and next-generation devices.

- As for research in the basic and advanced technology area, ETRI intensively carries out the study of core technologies in telecommunications, information, new materials and engineering, and physical sciences as a leading IT technology for the 21st century information society.

6.1.2 R&D Strategy

To realize ETRI's goal of becoming a world-class research institute and contributing to efficient development of core technology for the information society of the 21st century, ETRI is focusing on carrying out seven major strategies, as follows:

1. Research, develop, and implement strategic core technologies
2. Stress the qualities of IMPH (intelligent, multimedia, personal, and human) as research guidelines
3. Increase domestic and international joint cooperative research
4. Better coordinate the economic strategy of technology development
5. Encourage 3P (paper, patent, product) activities
6. Accumulate advanced technology information systems
7. Improve the R&D management system

6.1.3 Basic and Advanced Research

Basic research at ETRI is carried out in several selected areas of the physical, communications, information, materials, and engineering sciences. It challenges the limitations of the physical and engineering world and searches for new ideas, principles, and concepts to advance the telecommunications and information technologies for future generations.

In 1995, 50 staff members were conducting basic research at ETRI, of which about 40 hold PhD degrees. In 1994, over one hundred papers were submitted to major international journals and meetings, and about the same number to domestic journals. Staff have also filed more than 50 patents.

Since 1995, basic research at ETRI has been making important contributions to the technical and professional communities throughout the world, addressing fundamental issues that can impact the quality of human life in the coming years of the information age.

6.2. KOREA INSTITUTE OF SCIENCE AND TECHNOLOGY (KIST)

Founded on February 10, 1966, with funding from Korean President Pok and U.S. President Johnson, KIST was the first multidisciplinary research institute in Korea. Its areas of inquiry cover a broad range of scientific disciplines, as indicated by the organizational chart shown in Figure 6.1. KIST receives 70% of its budget from the government through MOST. KIST has over 800 employees, 270 of whom are PhDs, and has worked on over 6,000 projects in the last 30 years. Those projects relevant to the electronics field have included development of polyester film for video tape (SKC video tape now has a 40% world market share), diamond-like coatings for VCR heads, robotics for assembly, electroluminescent displays, optical fibers, and environmentally sound replacements for CFCs. In parallel with the government's future design of building a new Korea, KIST is committed to distinguishing itself as a revamped research institute dedicated to basic research and to the development of future-oriented, cutting-edge technologies.

KIST is now participating in one of the Highly Advanced National (HAN) projects called the KIST-2000 project, part of the larger "G-7" project designed to bring Korea up to the level of the Group of Seven countries, that is, up to world-class status. The KIST-2000 project is in accordance with the government decision encouraging the development of new technologies in applied sciences and engineering. These new technologies are expected to be the foundation for improving the Korean standard of living and also elevating industrial technology to more developed levels. KIST-2000 has five specific projects:

1.   Advanced Technology for Medical Science and Technology

2.   New Processing Technology of High-Tech Materials

3.   New Functional Devices for Information Technology

4.   Highly Intelligent Human Robot

5.   3-Dimensional Imaging Technology for Multimedia

Dr. Myung-Hwan Oh manages the KIST 2000 sub-program for "New Functional Devices for Information Technology," which has five specific projects:

- *Nanometer quantum devices for extra-high-speed switching*, to develop optoelectric device technology for devices of ultrahigh performance, aiming for terahertz speed and gigabit capacity with high efficiency, applicable to improving silicon and compound semiconductor device characteristics as well as to the micromachining field.

- *Micromagnetic devices for telecommunication power converters*, to develop miniaturized integrated electromagnetic devices in IC technology for telecommunications with good performance at frequencies up to 100 MHz. Development of new magnetic materials and micromachining processes for those magnetic materials are key goals.

- *Nonvolatile ferroelectric RAM (NVFRAM) devices for data processing*, to replace magnetic bubble memory and electrically erasable programmable read-only memory (EEPROM) with devices that are lower in power consumption and faster. Already under production are 4-16 kbit NVFRAM, and 256 kbit devices are under development.

- *Field emission devices (FED) for information display*, with the goal of producing a 20-in. full-color FED panel having high brightness and spatial resolution. In the first phase of the project, 3-in. monochromic FED are being produced. Some of the demonstrated FED prototypes are rather impressive.

- *Organic nonlinear optical materials for electro-optic devices*, considered the best candidates to give large optical non-linearity, transparency, ultra-fast response. and good processability in optoelectronic information processing applications. The goal of the research is to develop electro-optic modulators utilizing second order optical nonlinearity of the organic materials.

In general, KIST appears to operate on a par with many of the public research organizations in America. While post-doctorate and doctorate students are an integral part of the staff, KIST is not part of a university. The staff is involved with local universities, and through that interaction, they are able to get students to work on their projects. This is a great benefit to KIST, the students, and the Korean industry, as it gives valuable experience to the students prior to graduation.

6.3. KOREA ELECTRONICS TECHNOLOGY INSTITUTE (KETI)

The Korean Electronics Technology Institute (KETI) is an R&D institute focused on helping the development of small- and medium-sized domestic companies. The main R&D focuses are electronics technologies, including electronic component development (ASICs, sensors, optoelectronics) and system development (computer systems, digital systems, new media systems, and automobile control systems).

Foundation of an institute for electronics, communications, and information was suggested by the Electronics Industry Association in Korea (EIAK) in March 1989. The plan was forwarded MOTIE and selected as the core project of "Policy for Improving the Competitiveness in Manufacturing Industry" in 1991. KETI was finally established in Seoul in August 1991 and then relocated to a new site in PyungTaek, KyungGi-Do, a suburban area about one hundred miles south of Seoul in November 1993.

KETI is under MOTIE, which manages and regulates its R&D work. KETI makes its own plans and roadmaps with help of an advisory board that consists of experts from government agencies such as KIST, but KETI highly references U.S. and Japanese roadmaps.

KETI obtains approximately $400 million in funds for R&D expenditures each year; 50% of the funds is from MOIC, 30% from MOTIE, and the remaining 20% from MOST. There is also some small amount of contributions from domestic companies.

Under KETI, there is an R&D Division, a Planning & Coordination Development Division, and an ASIC Design Center. There are approximately 200 employees in KETI: 27% are PhDs, 36% have Master's degrees, 28% have bachelor's degrees, and 9% are technicians. The employees are about 50% experienced personal from private companies and about 50% university graduates. Most KETI research equipment is imported from the United States and Japan. KETI management hopes that in the future the equipment will be manufactured by Korean companies.

KETI's efforts are directed towards supporting small- to medium-sized companies in their technology development and competitiveness. The Industrial Doctor Program (IDP) and the Technical Membership Program (TMP) are currently being implemented to assist and solve bottleneck technologies faced by these companies. Research projects can be initiated by KETI or a company on a 50%-50% basis. Total funding for the projects initiated by KETI or the company is on a ratio of 70% funded by KETI to 30% funded by the business. KETI researchers work on-site at the companies and have performed 507 technical visits so far. The basic assistance to a company usually consists of marketing, project definition, and product design help. Product manufacture is subcontracted to outside companies. The funded companies are requested to pay back 50% of the research fund when a project begins and the remaining 50% when the project makes profits. KETI used to provide loans to small- and medium-sized companies, but no longer does so.

KETI funds universities such as Seoul National University and government organizations such as KIST for research projects. Approximately 30% of these projects are deliverable.

6.3.1 R&D

Although the purpose of KETI is to help small- and medium-sized companies, it has been establishing itself actively as a research organization. The main research focuses of KETI are in key components and materials technologies, information system development, and industrial instrument development. An overview of these research areas follows.

The key components and materials technology focuses include semiconductors, optoelectronic devices, and RF components. Current research is in progress to make the parts smaller, thinner, lighter, and more integrated. SMD-type high-frequency active/passive devices, HIC, and mixed devices such as transceivers and electronic circuits are being developed. The key materials are ceramic bulk materials, thin films, and single crystals. Research emphasis on ceramic materials includes powder processing, characterization, and device implementation on various active/passive devices. Research emphasis on thin film material includes the development of thin film materials used in various sensors and semiconductors, wide-band gap materials, and piezoelectric and magnetic thin film materials. Research emphasis on single crystals includes growing, processing, characterization, and implementation of single crystals on devices such as optical isolators.

An additional research emphasis within the key component and materials category is the Electro-21 Project. The project includes 18 subprojects for core technology development, 200 subprojects for standardization, and 1,500 subprojects for bottleneck technologies of small- to medium-sized companies.

The information system development focuses are the computer, software, cable TV, and multimedia. KETI has set up a 5-year Development Plan for Information Technology (1992-1996) to build up the domestic information industry. In the computer area, KETI is localizing notebook PCs and developing next-generation pen-computers, palmtop computers, and PDAs. In the cable TV area, KETI developed a cable TV subscriber management system and network management system for the Korean CATV service. The next research emphasis will be on the development of core technologies for digital TV technology and multimedia service systems. In the multimedia area, KETI is conducting research on signal processing technology, as well as on the design of key components such as MPEG chips and the development of GSM and ADC digital cellular phones.

KETI's industrial instrument development focus includes industrial electric equipment such as control motors, controllers, SMPS, power systems for communications, and general purpose AC drivers. A factory automation group in KETI has developed several kinds of automatic inspection and assembly equipment. The group is currently working on developing an Integrated Manufacturing System (IMS).

The major objective of the ASIC Design Center is to provide small- and medium-sized domestic companies with "total solutions to ASIC design," consisting of design environments, designers, and research funds. The center plans to carry out joint design programs involving industries, universities, and research institutes. The major activities of the Center are in the development of DSP chip set for mobile telecommunication, development of chip sets for multimedia, development of general purpose chips such as analog to digital converters, support of ASIC design for small- and medium-sized companies such as LED, FAX, and LBP (laser beam printer) controllers, and card readers, mixed signal chip design, and general top-down design using high-level synthesis for DSP chips and chip sets for TDMA and multimedia.

6.4. KOREA ADVANCED INSTITUTE OF SCIENCE AND TECHNOLOGY (KAIST)

In the early 1970s, the government of the Republic of Korea realized the need for the country to produce highly qualified scientists and researchers who could lead the science and technology industry. The Korean Advanced Institute of Science (KAIS), the first government-supported academic institute, was established in 1971. It grew steadily for ten years, and then in 1981, the institute merged with the Korean Institute of Science and Technology (KIST), a product-oriented research institute, thus producing KAIST. However, after the merger, the new institute apparently lost its effectiveness in handling its two main functions (academic support and R&D). So in 1989, KIST and KAIS were separated again, and KAIS merged with the Korean Institute of Technology, a government-funded undergraduate college for science and engineering. This new academic institution retained the name KAIST.

The Accreditation Board for Engineering & Technology (ABET, a U.S. organization for evaluating institutes of higher education in science and engineering) has evaluated KAIST and given it high marks for both graduate and undergraduate programs. PhD candidates at KAIST (approximately 600) are exempt from otherwise mandatory military service.

6.4.1 Branches

KAIST's main campus is located in Taedok Science Town, Taejon, the center of the Korea's most advanced science and technology activities. The campus contains 6 schools, 20 departments, and 5 affiliated research institutes.

The Seoul campus has four departments: the Departments of Management Information Systems, Information and Communication, Automation and Design Technology, and Management and Information Systems. KAIST has a total student population of 2,500 undergraduates and 3,000 graduates, and it has 800 faculty and staff.

6.4.2 Electrical Engineering Department

The Department of Electrical Engineering at KAIST, aiming at being an advanced institute for excellence in science and technology training, is one of the largest departments in Korea, with modern facilities for teaching and research in electrical engineering, including communications, information, semiconductor, computer, microwave, control, and system engineering. The department offers BS, MS, and PhD degrees. It has 39 faculty members and the largest student body of any department at KAIST. Department objectives include teaching with modern concepts, leading-edge research, and service to society by supplying highly qualified engineers and scientists capable of solving real-world problems. Department curriculum offers a large number of courses from basics to practical applications, so that the student may have extensive academic background and gain experience in seminars and experimental work. The R&D activities cover so-called high-tech-related works in electrical engineering in close collaboration with industries and research institutes.

Three research centers, the Satellite Technology and Research Center funded by the Korea Science and Engineering Foundation (KOSEF), the Center for Industrial Electronics Technology supported by MOTIE, and the Center for High-Speed Integrated Circuits funded by KOSEF, have been established at the Electrical Engineering Department to promote interdisciplinary research and collaboration with industries and with other research institutes. The undergraduate program offers basic courses dealing with circuits and systems, semiconductors, telecommunications, computer and control systems, and includes several laboratories. The graduate programs focus on thesis research work in close collaboration with domestic industries.

Relevant research areas are classified into five specialty areas:

1. *Computer and circuit design,* focused on developing practical computer architecture and application software in the fields of digital computers, VLSI design, and neural computing

2. *Communications and signal processing,* focused on enhancing fundamental understanding of and developing core technologies for processing, transmission, storage, and display of various types of data, including audio, video, and graphic signals.

3. *Microwave and electro-optics:* the microwave group is focused on detecting underground obstacles and the algorithms of the imaging radars, as well as on RCS analysis, EMI/EMC analysis, and related software development; the electro-optics group is focused on experimental and theoretical studies of optical devices and systems for optical communications, optical sensing, and optical information processing.

4. *Semiconductors,* focused on all aspects of fabrication processes, device modeling and design, and circuit applications for discrete and integrated semiconductor devices based on various materials such as silicon, GaAs, and HgCdTe.

5. *Control and systems,* focused on theories and applications of control systems, robotics, and control elements that are essential to production lines and process control in industry, including IC design of control circuits and driving systems.

One main focus of the Electrical Engineering Department is on semiconductor education, research, and infrastructure for nonmemory devices. Nonmemory technologies require a new focus on design and technology concepts, where previously, memory work primarily was focused on manufacturing. Success in the nonmemory area will require more teamwork, long-range planning, theory, cross-disciplinary interaction, and product focus — areas in which Korean technologists have little experience. KAIST and other Korean universities have identified these barriers and have instituted programs to address them.

6.5. SEOUL NATIONAL UNIVERSITY

Seoul National University (SNU) is the premier university in Korea. The scope of academic disciplines taught at SNU cover the entire range of sciences, arts, humanities, and engineering. Rated at the top of the public university system, SNU has traditionally attracted the top students in all fields of study, and the electrical engineering program is no exception. SNU is primarily funded by the Ministry of Education; however, in recognition of the importance of encouraging high-tech research, exceptions have been made to some of the Ministry's more restrictive policies to allow certain parts of the university to operate quite independently.

The main attraction of interest to the electronics industry is the Inter-university Semiconductor Research Center (ISRC) based at SNU. This center provides for research projects and education for future semiconductor engineers. The center also maintains a first-rate teaching semiconductor fabrication facility that is equivalent to similar facilities at MIT, the University of Illinois, and the University of California at Berkeley. This facility regularly fabricates multiple-project chips (MPC) as a service to other universities in Korea. At the time of the WTEC team's visit to Korea, nearly 300 designs had been fabricated, 10% of which were designed at companies and 70% of which were designed at other Korean universities.

The level of facility development at SNU is not quite at the level of Stanford or Cornell, which have intimate ties to U.S. equipment manufacturing companies. Korea does not have an indigenous equipment manufacture infrastructure to support development of new processing facilities, and the government alone cannot be expected to fund this sort of effort on a large scale; nevertheless, the equipment that is on-line at SNU is still impressive by any American standard. Most impressive is the routine use of a high-quality electron-beam (Cambridge EBMF) tool for deep submicron mask making and direct-write lithography. Cornell is the only U.S. school with such a capability.

The ISRC is funded at a level exceeding U.S. $5 million annually, with just over half of the money coming from industry. Projects include novel circuit design, device modeling, computer-aided design, signal processing, neural networks, novel semiconductor process technologies, and many others. The university has provided the Center with the ability to perform proprietary research directly funded by the industry, where the intellectual property would be owned by the funding company.

The industrial funding at SNU is recognized to be primarily a recruitment method for the companies to attract top students from the top university in the country. Although proprietary arrangements are made, the results of research tend to be a well-qualified graduate rather than important discoveries that are considered critical to a company's competitive strategy. The education at SNU is considered to be second to none in Korea, so top students graduating from the electronics program are highly sought after by the industry.

The Center also provides for 2 week-long short courses in the areas of semiconductor processing, VLSI and ASIC design. These courses are offered 6 times a year, with enrollment of approximately 20 students per class. The enrollment consists of graduate, undergraduate, and continuing education students who are not pursuing a degree. ISRC also offers a full range of design and architecture courses for full-time students to major in VLSI design and in semiconductor processes.

6.6. POHANG UNIVERSITY OF SCIENCE AND TECHNOLOGY (POSTECH)

In the past thirty years, Korea has been driven by a tenacious determination to stand independently as a nation of international prominence. As Korea's rapid industrialization and modernization process has unfolded, the need has become evident for increased self-reliance in science and technology to launch Korea into the forefront of the high technology era.

Recognizing the national need for technological advancement, the Pohang Iron and Steel Company, Ltd. (POSCO), the second largest steel producer in the world, decided to build a world-class university of science and technology in the city of Pohang, 210 miles southeast of Seoul. By doing so, POSCO could also answer its own problematic R&D requirements. As a result, the Pohang University of Science and Technology was launched in 1986 as Korea's first private research-oriented institution of higher learning in science and technology.

Since its inception, the commitment to the goals of a quality education, progressive research, and attention to real-world problems have remained the hallmarks of POSTECH's mission. The institution's aim is to educate Korea's gifted youths to become national leaders as outstanding scientists, engineers, and researchers; to further lead Korea to the pinnacle of the high-tech era by serving as a link between research, education, and industry; and to contribute to the technological development of the world.

The founders of POSTECH planned out a detailed agenda to accomplish their mission. This agenda includes a philosophy of selecting a small number of elite students from among Korea's most promising youths and educating them in an "ideal" environment. This "ideal" environment includes an excellent faculty comprised of scholars selected from renowned institutions around the world. Keeping the ratio of students to faculty low ensures that each student receives the maximum guidance in his or her academic endeavors. POSTECH also provides its scholars and students with the finest facilities and the most advanced equipment to facilitate their success. At present, 2,100 students (1,200 undergraduates and 900 graduates) and 200 faculty members are carrying out the mission in four departments of science, six departments of engineering, six research centers of excellence, and many other research laboratories, including the Pohang Accelerator Laboratory, the first particle accelerator in Korea.

The successful opening of the Pohang University of Science and Technology marked a turning point in the emphasis of scientific and technical education in Korea from quantity to quality, from mass education to the fusion of teaching and research, and from general theory to attention to real-world problems. Now a leading university in Korea, POSTECH aims to be a world-class university in science and technology in the future.

6.7. SUMMARY

It is through its educational system that the Republic of Korea is building up an intellectual infrastructure and preparing itself for long-term success. As excellent as the university system is in Korea, professors still feel highly motivated to reform their curricula. They feel that creativity needs to be better encouraged to a level closer to that of the West.

Korean university-level educators have great respect for the U.S. educational system. This is evidenced by the overwhelming majority of

U.S.-educated faculty in Korea. Many of the schools, including KAIST, POSTECH, and IAE, have official collaborations with universities in the United States.

Chapter 7

A NEW GLOBAL FORCE IN ELECTRONICS

The Republic of Korea has catapulted itself into the role of a leading player in the global electronics industry. Korean electronics companies are now the major DRAM suppliers in the world. Since 1991, semiconductor sales have increased sevenfold. Furthermore, the Korean electronics industry is becoming a significant force in display technologies and electronics products to advance information technologies.

The success of the Korean electronics industry is generally attributed by the nation's four leading chaebols to the foresight of company founders and to the sound strategy and aggressive and astute work of their executives. In preparing for the future, many Korean companies are putting up to 4% of total sales into tax-sheltered "reserve funds" for investment in areas like R&D, employee development, and plant improvement. In addition, through government tax programs, firms are deducting up to 15% of total expenditures on training and manpower development. Ten percent of R&D facility construction cost is tax-deductible, and R&D and test facilities can be depreciated at 90% per year.

Government assistance and support has played a significant role in preparing the Korean electronics industry for the 21st Century. This assistance is now focusing on new priorities, such as products to support information technologies and the startup of a domestic semiconductor equipment industry. As part of a broad national outline for technological development, the Korean government continues to nurture its electronics industry through legislating supportive tax incentives, banking regulations, and incorporation laws, and through direct financial support from ministries and universities. The government has considerably lowered the cost of basic research and plant modernization, while ensuring that universities educate skilled technologists and maintain a pool of basic researchers and educators who act as consultants to industry. Considerable planning and coordination has minimized redundancy of effort and synchronized development in many related areas. Government also supports development of science and technology parks and works internationally in various capacities to help nurture development of a highly sophisticated technology infrastructure. Its efforts in support of the domestic electronics industry have been remarkably successful, although there is a certain amount of resistance in the industry to government involvement, especially in the most highly competitive and mature technology fields.

Korea's public and private (industry-supported) universities and institutes are an integral part of the striking and rapid growth of the country's electronics industry. Besides working energetically to continually upgrade the quality of training for technicians and engineers and making R&D contributions, educators are clearly focused on developing new creative, independent, and cooperative thinking skills within the population that will support the national commitment to be self-reliant in high-technology innovation as a means to improving economic strength.

The development plans of the Samsung, LG, Hyundai, and Daewoo chaebols still appear to dominate national technology policy. As these chaebols grow ever larger and develop vested interests in maintaining the status quo, their influence may lead to some difficulties for emerging companies developing non-standard technologies. The Korean government is attempting to assist small and emerging companies through organizations like the Korean Electronics Technology Institute, set up by the Ministry of Trade, Industry, and Energy. However, even here emphasis appears to be on more mainstream silicon-based technologies, in line with the development plans of the four largest, established Korean electronics firms.

Korea has built and is continuing to build a "stand-alone" capability in a broad range of electronics technologies. These include DRAM, SRAM, and ASIC design approaches; electronic materials and packaging; and development of new products, especially those such as displays that are fundamental to the information services market. The nation's strategic focus is on achieving dominance not only in production and manufacture of electronics products and components, but also in creation and innovation of new technologies in the field. Korea is determined to remain internationally competitive in electronics in the long run and is prepared to commit the required long-term financial and logistical resources to achieve its goals.

REFERENCES

Bae, M. 1987. The Korean semiconductor industry: a brief history and perspective. *Solid State Technology*. 30(October):141-144.

Bark, T.H. 1995. Korea-U.S. economic relations and the new economic order. *Nation's Business*. 83(July):56-68.

Bennett, E. 1986. Korea keeps investing in chips despite downturn. *Electronics*. 59(April):21.

Berger, M. 1987. Korea aims for the top in VLSI by 1991. *Electronics*. 60(April):44.

Berger, M. Korean chip makers aren't backing off. *Electronics*. 58(December):28-29.

Bottoms, D.T. 1993. Korea DRAMS duty free! *Electronics*. 66(March):1.

Byun, B.M. 1994. The growth and recent development of the Korean semiconductor industry. *Asian Survey*. 34(8):706.

Byun, B.Y. and Ahn, B.H. 1989. Growth of the Korean semiconductor industry and its competitive strategy in the world market. *Technovation*. 9:635.

Castells, M. and Hall, P. 1994. *Technopoles of the World: The Making of Twenty-First-Century Industrial Complexes*. London:Routledge.

Choi, H.B. 1995. Conversation with authors. Seoul:Hyundai Electronics Industries.

Choi, S.H. 1995. Conversation with authors. Seoul:Hyundai Electronics Industries.

Claveloux, D. 1993. EC to investigate dumping of electrolytic capacitors from Korea and Taiwan. *Electronics*. 66(March):3.

Connelly, M.J. 1995. Korea: Remembering America's Forgotten War. *Washington Post*. (July 25):A17-21.

DOC (U.S. Department of Commerce). 1995. *The Big Emerging Markets. 1996 Outlook and Source Book*. Lanham, MD: Bernham Press. 260-283.

_____. 1996. *Globalizing Research and Development: Methods of Technology Transfer Employed by the Korean Public and Private Sector*. Washington D.C.: Office of Technology Policy, Technology Administration, DOC.

Eckert, C. J., et al. 1990. *Korea Old and New. A History*. Cambridge, MA: Ilchokak Publishers (Korea Institute, Harvard University).

Electronic Industries Association of Korea. 1995. *Korea Electronics Parts.* Seoul: Electronic Industries Association of Korea.

Electronic and Telecommunications Research Institute. 1995. *Annual Report 1994.* Seoul:ETRI.

Electronic News. 1995. Hyundai, AirWave Pact Targets PCS. 41(July).

Elsevier Science Ltd. 1995. New DRAM fabs for Oregon. *Integrated Circuits International.* 19(July):8.

Heyler, D.A. 1993. Semiconductor industry status report: South Korea, Taiwan, Singapore. *Solid State Technology.* 36(November):29-30.

Hyundai Electronics Industries. 1995. *My Perspective.* Seoul:Hyundai Electronics Industries.

Holstein, W.J. and Nakarmi, L. 1995. Korea. *Business Week.* (July 31):56-64.

Kim, N. H. 1995. EDO DRAM standard gains support. *Electronics.* 68(January):6.

_____. 1995. Top 10 Dominate Memory Market. *Electronics.* 68(January):7.

_____. 1995. LG Electronics develops MPEG-2 decoder for HDTV. *Electronics.* 68(January):6.

_____. 1995. Korea chip makers capture nearly one-quarter of world market. *Electronics.* 68(January):7.

_____. 1994. Korea's semicon industry will continue strong growth. *Electronics.* 67(March):6.

_____. 1994. South Korea still relies heavily on foreign chip equipment and materials. *Electronics.* 67(May):6.

_____. 1994. Samsung claims world's first 256M DRAM. *Electronics.* 67(September):12.

_____. 1994. Korean TFT-LCD makers are ready to challenge Japan. *Electronics.* 67(May).

_____. 1994. Korea's semicon exports will top US$10 million. *Electronics.* 67(January).

_____. 1994. Korea launches fostering program for chip production equipment. *Electronics.* 67(March):6.

_____. 1994a. Korea expects strong growth. *Electronics.* 67(January).

_____. 1994. Goldstar to enter RDRAM arena. *Electronics.* 67(April):12.

_____. 1994. Daewoo re-enters semiconductor fray, wants capacity to match country's best. *Electronics.* 67(April):5.

_____. 1994. Chip sales boom in South Korea. *Electronics.* 67(July):4.

_____. 1993. Korean SRAM sales rise sharply. *Electronics.* 66(July):5.

_____. 1993. U.S. rejects Korean proposal to head off DRAM dumping penalties. *Electronics*. 66(March):13.

_____. 1993. S. Korea will begin enforcing semiconductor copyright law. *Electronics*. 66(June):4.

_____. 1993. Korean semicon industry tops US $1 billion in Q1 1993. *Electronics*. 66(July):14.

_____. 1993. Korean semicon giants race to produce 16M DRAM, flash memory. *Electronics*. 66(January).

_____. 1993. Korean LCD makers call for government help. *Electronics*. 66(September):6.

_____. 1993. Korea plays catch-up in semicon production equipment market. *Electronics*. 66(July):7.

_____. 1993. Korea nears self-sufficiency in semiconductor materials. *Electronics*. 66(May):1.

_____. 1993. Dumping duties won't slow Korean DRAM shipments, says trade official. *Electronics*. 66(May):12.

_____. 1993. Anam launches commercial production of 225-pin ball grid array package. *Electronics*. 66(September):3.

_____. 1993 Korea's electronics giants face off in TFT LCD market. *Electronics*. 66(August).

_____. 1993 Korea's 256M DRAM project begins in November. *Electronics*. 66(October):14.

_____. 1992. Koreans seek suspension agreement during DRAM dumping investigation. *Electronics*. 65(November):14.

_____. 1992. Korea's chip makers battle over 64M DARM announcement. *Electronics*. 65(October):4.

_____. 1992. EC slaps tax on Korean DRAMS. *Electronics*. 65(September):3.

Kim, N.H., and Shandle, J. 1992 ITC ruling may spark U.S./Korea trade talks. *Electronics*. 65(June):1.

KETI (Korea Electronics Technology Institute). 1994. *Chonja Kisul Yechuk (Electronics Technology Forecasting)*. Seoul: KETI. (November):267-341. Translated at CALCE-EPRC University of Maryland.

Kuk, Y. 1995. Conversation with authors. Seoul:Seoul National University.

Kyung, C.M. 1995 Conversation with authors. Taedok:Korea Advanced Institute of Technology

Lee, D.H. 1995. Conversation with authors. Seoul:Hyundai Electronics Industries.

Lee, K.P., et.al. 1995. A Process Technology For 1 Gigabit DRAM. *Proceedings IEEE IEDM.* IEEE cat. no. 95CH35810, p. 907.

LG Electronics Co. 1995. *Annual Report.* Seoul: LG Electronics Co.

McLeod, J. 1994. Korea aspires to lead in high technology. *Electronics.* V(June).

Ministry of Science and Technology. 1995. *Science And Technology In Korea.* Seoul:MOST.

Ministry of Science and Technology. 1994. *Statistics of International Cooperation, Republic of Korea.* Seoul:MOST.

Ministry of Science and Technology. 1995. *Highly Advanced National Project: A Plan Towards the 21st Century.* Seoul:MOST.

Nahm, A.C. 1993. *Introduction to Korean History and Culture.* Seoul:Hollym Press.

Noh, M.R. 1995 Conversation with authors. Seoul:IC Design Education Center.

OECD. 1996. *Reviews of National Science and Technology Policy: Republic of Korea.* OECD 81-103.

Oh, M.H., Kim, Lee. 1995. Conversation with J. Budnaitis. Seoul:KIST.

Park, C.S. 1995. Conversation with authors. Seoul:Hyundai Electronics Industries.

Park, J.K., et.al. 1995 Isolation merged bit line cell (IMBC) for 1 Gb DRAM and beyond. *Proceedings IEEE IEDM.* IEEE cat. no. 95CH35810, 907.

Pettit, F. 1989 *The Pohang Iron and Steel Company: Its research institute and technical university in South Korea.* ONRFE.

POSTECH. 1996. *Prospectus.* Pohang:POSTECH.

Rosenblatt, A., T. Perry, T. Dambrot, P. Gwynne, G. Watson, and K. Fitzgerald. 1991. Formula for competitiveness. *IEEE Spectrum.* 28(June).

Scientific Information Bulletin. 14(4):51-64.

Swinbanks, D. 1993. What road ahead for Korean science and technology? *Nature.* 364(July):377-384.

Wang, K.W. 1995. Conversation with authors. Seoul:Inter-University Semiconductor Research Center.

Internet References

Daewoo Electronics Home Page. http://parrot.dwe.co.kr/

ETRI Home Page. http://mouth.etri.kr/

Hyundai Home Page. http://www.hit.co.kr/

KAIST Home Page. http://camis.kaist.ac.kr:8080/Welcome.en.html

KIST Home Page. http://news.kreonet.re.kr/kistDir/kisthome.html

Samsung Electronics Co. Home Page. http://www.sec.samsung.co.kr/

WWW Servers in Korea, http://www.dongguk.ac.kr/

WWW servers in Korea,
  http://flower.comeng.chungnam.ac.kr/sharon/www-server-in-korea-eng.html

LIST OF ACRONYMS

ADC	Analog-digital converter
ADM	Add/drop multiplexer
AIN	Advanced intelligent network
ALE	Atomic layer epitaxy
APCVD	Atmospheric pressure CVD
ASIC	Application-specific integrated circuit
ASIS	Add/drop signal interfacing subsystem
ATM	Atomic force microscope
ATM	Asynchronous transfer mode
B/W	Black and white cathode ray tube
BGA	Ball grid array
BIOS	Basic input/output system
B-ISDN	Broadband integrated services digital networks
BOSS	BDCS operation & surveillance subsystem
B-TA	Broadband-terminal adapter
CAD	Computer-assisted design
CATV	Cable television
CBIC	Cell-based integrated circuits cell library
CCDs	charge-couple devices
CCS	Common channel signaling
CDMA	Code division multiple access
C-DRAM	Cache DRAM
CD-ROM	Compact disk read-only memory
CFCs	Chlorofluorocarbons
CIM	Computer-integrated manufacturing
CMOS	Complementary metal-oxide semiconductors
CPU	Central processing unit
CR	Change request
CRT	Cathode ray tube
CSP	Chip scale package
CW	Continuous wave
DARI	Data application for reliability information
DIXS	Digital cross-connect subsystem
DRAM	Dynamic random access memory
DSP	Digital signal processing
DVT	Desktop videoconference terminals
EDIRAK	Electronics Display Industrial Research Association of Korea
EEPROM	Erasable electrically programmable read-only memory
EL	Edge lit

EMI	Electromagnetic interference
EPROM	Erasable programmable read-only memory
ERIS	ETRI Reliability Information System
ETRI	Electronics and Telecommunications Research Inst.
FED	Field emission display
FPD	Flat panel display
FPGA	Field programmable gate array
FPLMTS	Future public land mobile telecommunications systems
FRAM	Ferroelectric RAM
FTC	Fiber to the curb
GaAs MMIC	Gallium arsenide monolithic microwave IC
GATT	General Agreement on Tariffs & Trade
GIANT	Gigabit information processing and networking technology
GPS	Global positioning system/satellite
HAN	Highly Advanced National Project, coordinated by KAITECH
HBT	Heterojunction bipolar transistor
HDTV	High definition television
HEI	Hyundai Electronics Industries Co., Ltd.,
HIC	Hybrid IC
I/O	Input/output
IAA	Industrial Advancement Administration of MOTIE
IC	Integrated circuit
ICPS	Information communications processing system
ICPS	Information Communications Processing System
IMBC	Isolation merged bitline cell
IMPH	Intelligent, multimedia, personal, and human
IN	Intelligent network
ISAPs	Information service access points
ISDN	Integrated services digital network
ISO	International Standards Organization
ISRC	SNU's Interuniversity Semiconductor Research Center
IT	Information technology
KAIST	Korea Advanced Institute of Science and Technology
KAITECH	Korea Academy of Industrial Technology
KARI	Korea Aerospace Research Institute
KEPCO	Korean Electric Power Corporation
KETI	Korean Electronics Technology Institute
KETRI	Korea Electric Technology Research Institute
KIER	Korea Institute of Energy Research
KIST	Korea Institute of Science and Technology
KORDIC	Korea Research and Development Information Center
KOSEF	Korean Science and Engineering Foundation
KRISS	Korea Research Institute of Standards and Science

KSIA	South Korean Semiconductor Industry Association
KTB	Korea Technology Banking Corporation
LAN	Local area network
LC	Liquid crystal
LCD	Liquid crystal display
LOC	Lead on chip
LOCOS	Local oxidation of silicon
MCU	Micro controller unit
MEMS	Microelectromechanical systems
MESFETs	Metal-semiconductor field-effect transistors
MMIC	Monolithic microwave IC
M-OAMs	Mediation-operation administration and maintenance systems
MOCVD	Metal oxide chemical vapor deposition
MOIC	Ministry of Information and Communication
MOS	Metal oxide semiconductor
MOST	Ministry of Science and Technology
MOTIE	Ministry of Trade, Industry, and Energy
MPC	Multiple-project chips
MUX/DMUX	Multiplexer/demultiplexer
NICS	Network information control system
NII	National information infrastructure
NVFRAM	Nonvolatile ferroelectric random access memory
OEM	Original equipment manufacture
PBH	Planar buried heterostructures
PCS	Personal cellular system
PCS	Personal communications services
PDA	Personal data assistant
PDH (DS3)	Parallel digital hierarchy
PDIP	Plastic dual in-line package
PDP	Plasma display panel
P-HEMT	Pseudomorphic high-electron mobility transistors
PLCC	Plastic leaded chip carrier
PLD	Programmable logic device
PLS	Pohang Light Source (accelerator)
POSTECH	Pohang University of Science and Technology
PPM	Parts per million
PSDN	Public switching data network
PSTN	Public switching telephone networks
PTP	Point-to-point
QFP	Quad flat pack
QSOP	Quarter-size outline package
R-DRAM	Rambus DRAM
RF	Radio frequency
RTD	Resonant tunneling diode

Rx MMIC	Receiver MMIC
SDH	Synchronous digital hierarchy
S-DRAM	Synchronous DRAM
SERI	System Engineering Research Institute
SHM	Self healing mesh
SHR	Self healing ring
SIA	Semiconductor Industry Association (U.S.)
SiGe HBT	Silicon germanium heterojunction bipolar transistor
SIGNOS	Signaling network operations/management system
SMD	Surface mount device
SMOT	Synchronous multiplexer & optical terminal
SMQW	Strained multi-quantum well
SNU	Seoul National University
SOJ	Small outline "J" form
SOP	Small outline package
SRAM	Electronic static random access memory
S-SEEDs	Symmetric-self electro-optic effect devices
SSMBE	Solid source molecular beam epitaxy
SSOP	Shrink small outline package
STEPI	Science and Technology Policy Institute (MOST)
STM	Scanning tunneling microscope
STM-N	Synchronous transport module, level N (N = 4.16.64)
STN	Super twisted nematic (type of LCD)
TA	Terminal adapter
TDM	Time-division multiplexing
TDMA	Time division multiple access
TFT	Thin film transistor
TQFP	Thin quad flat pack
TQM	Total quality management
TSOP	Thin small outline package
TSSOP	Thin shrink small outline package
TV	Television
UHVCVD	Ultrahigh vacuum chemical vapor deposition
ULSI	Ultra large-scale integrated circuits
UPT	Universal personal telecommunications
VAN	Value added network
VFD	Vacuum fluorescent display
VGA	Video graphics array
VLSI	Very large-scale integrated circuits
VOD	Video-on-demand

INDEX